AF361866

Recent Advancement of Thermal Fluid Engineering in the Supercritical CO$_2$ Power Cycle

Editors

Jeong Ik Lee
David Sánchez

MDPI • Basel • Beijing • Wuhan • Barcelona • Belgrade • Manchester • Tokyo • Cluj • Tianjin

Editors
Jeong Ik Lee
Korea Advanced Institute of Science & Technology
Korea

David Sánchez
University of Seville
Spain

Editorial Office
MDPI
St. Alban-Anlage 66
4052 Basel, Switzerland

This is a reprint of articles from the Special Issue published online in the open access journal *Applied Sciences* (ISSN 2076-3417) (available at: https://www.mdpi.com/journal/applsci/special_issues/Supercritical_CO2_Power_Cycle).

For citation purposes, cite each article independently as indicated on the article page online and as indicated below:

LastName, A.A.; LastName, B.B.; LastName, C.C. Article Title. *Journal Name* **Year**, *Article Number*, Page Range.

ISBN 978-3-03943-016-1 (Hbk)
ISBN 978-3-03943-017-8 (PDF)

Contents

About the Editors

Jeong Ik Lee was educated at Seoul National University at the Department of Nuclear Engineering until 2003. In 2007, he completed his Ph.D. degree in Nuclear Science and Engineering from Massachusetts Institute of Technology with support from the MIT Manson Benedict Fellowship. In 2010, he took up an assistant professorship in KAIST and moved to UAE to join Khalifa University of Science and Technology as a Joint-Professor in the Nuclear Engineering program. After staying in UAE for one year, he came back to KAIST and was promoted to Associate Professor in 2015. In 2017, he stayed at the University of Cambridge as a visiting scholar for one year and was involved in developing innovative nuclear systems. His research group in KAIST is very active in the area of supercritical CO_2 power cycle technology development. His research group is not only working on nuclear energy systems but also expanding the application area of the technology to the waste heat recovery systems for gas turbines and fuel cells. Research on the supercritical CO_2-cooled micro modular nuclear reactor was recognized as the top ten research achievements of KAIST in 2014. He collaborated with domestic and international companies to develop an appropriate waste heat recovery system with the supercritical CO_2 power system technology.

David Sánchez is a Full Professor of Energy Systems and Turbomachinery at the Department of Energy Engineering of the University of Seville. At this institution, he obtained his MSc in Mechanical Engineering in 2001 and, in 2005, he completed a Ph.D. working on the smart integration of high-temperature fuel cells and bottoming heat engines such as micro gas turbines, Stirling engines or supercritical CO_2 power cycles. At the same time, he expanded his postgraduate education with a specialization in the area of gas turbines at international institutions like Cranfield University (United Kingdom) and the Von Karman Institute for Fluid Dynamics (Belgium). Throughout his career, he has provided support to the Concentrated Solar Power industry under development in Spain and worldwide. His research interests remain focused on gas turbines for power generation applications, with a focus on micro gas turbine systems, he has been very active in the development of supercritical CO_2 technologies for concentrated solar power applications in the last decade. This research interest has, more recently, expanded into the combined production of power and water, along with the growing concern about the water–energy nexus raised by the scientific communities and governments. These activities are funded through national and international R&D programs and also through industrial sponsorship.

Editorial

Recent Advancement of Thermal Fluid Engineering in the Supercritical CO_2 Power Cycle

Jeong Ik Lee [1],* and David Sanchez [2]

[1] Department of Nuclear and Quantum Engineering, Korea Advanced Institute of Science and Technology, 291 Daehak-ro, Yuseong-gu, Daejeon 34141, Korea
[2] Department of Energy Engineering, University of Seville, Camino de los Descubrimientos s/n, 41092 Seville, Spain; ds@us.es
* Correspondence: jeongiklee@kaist.ac.kr

Received: 29 July 2020; Accepted: 31 July 2020; Published: 3 August 2020

The supercritical CO_2 (S-CO_2) power cycle is an emerging energy technology that has potential to revolutionize the conversion process of heat to mechanical or electric power. Currently, the technology development is being actively pursued in many countries thanks to the support of governments and the industry. At the same time, the technology is already being commercialized in the waste heat recovery sector successfully and it is diffusing to other conventional energy source applications such as gas, coal and nuclear power. For renewable energy sources such as concentrated solar power, the S-CO_2 power cycle stems as a technology which can enable a substantial reduction of the cost of electricity, thus contributing to a larger penetration of environmentally friendly, dispatchable and cost-effective energy technologies.

This Special Issue contains up-to-date techno-economical information regarding the S-CO_2 power cycle. The contents of this issue cover from component level technologies such as turbine [1–3], compressor [4] and heat exchanger [5] to system level information such as cycle analysis [6,7] and economic assessment [8] of the S-CO_2 power cycle. The articles in the issue were provided by groups of researchers spread across globally and they come from different types of organizations, which also tells how active this area is being researched at the moment. The editors would like to thank all the authors who contributed to this Special Issue and feel very privileged to have had the opportunity to produce this Special Issue. They also hope that this Special Issue contributes to the advancement of the S-CO_2 power cycle technology by informing and inspiring many researchers in this field.

Author Contributions: All authors contributed equally to the preparation of this manuscript. All authors have read and agreed to the published version of the manuscript.

Funding: This research received no external funding.

Acknowledgments: This publication was only possible with the invaluable contributions from the authors, reviewers, and the editorial team of Applied Sciences.

Conflicts of Interest: The authors declare no conflict of interest.

References

1. Salah, S.I.; Khader, M.A.; White, M.T.; Sayma, A.I. Mean-Line Design of a Supercritical CO_2 Micro Axial Turbine. *Appl. Sci.* **2020**, *10*, 5069. [CrossRef]
2. Samad, T.E.; Teixeira, J.A.; Oakey, J. Investigation of a Radial Turbine Design for a Utility-Scale Supercritical CO_2 Power Cycle. *Appl. Sci.* **2020**, *10*, 4168. [CrossRef]
3. Wang, Y.; Li, J.; Zhang, D.; Xie, Y. Numerical Investigation on Aerodynamic Performance of SCO_2 and Air Radial-Inflow Turbines with Different Solidity Structures. *Appl. Sci.* **2020**, *10*, 2087. [CrossRef]
4. Shi, D.; Xie, Y. Aerodynamic Optimization Design of a 150 kW High Performance Supercritical Carbon Dioxide Centrifugal Compressor without a High Speed Requirement. *Appl. Sci.* **2020**, *10*, 2093. [CrossRef]

5. Seo, H.; Cha, J.E.; Kim, J.; Sah, I.; Kim, Y.-W. Design and Performance Analysis of a Supercritical Carbon Dioxide Heat Exchanger. *Appl. Sci.* **2020**, *10*, 4545. [CrossRef]
6. Salim, M.S.; Saeed, M.; Kim, M.-H. Performance Analysis of the Supercritical Carbon Dioxide Re-compression Brayton Cycle. *Appl. Sci.* **2020**, *10*, 1129. [CrossRef]
7. Ham, J.K.; Kim, M.S.; Oh, B.S.; Son, S.; Lee, J.; Lee, J.I. A Supercritical CO_2 Waste Heat Recovery System Design for a Diesel Generator for Nuclear Power Plant Application. *Appl. Sci.* **2019**, *9*, 5382. [CrossRef]
8. Crespi, F.; Sánchez, D.; Martínez, G.S.; Sánchez-Lencero, T.; Jiménez-Espadafor, F. Potential of Supercritical Carbon Dioxide Power Cycles to Reduce the Levelised Cost of Electricity of Contemporary Concentrated Solar Power Plants. *Appl. Sci.* **2020**, *10*, 5049. [CrossRef]

Article

Mean-Line Design of a Supercritical CO_2 Micro Axial Turbine

Salma I. Salah *, Mahmoud A. Khader, Martin T. White and Abdulnaser I. Sayma

School of Mathematics, Computer Science and Engineering, City University of London, Northampton Square, London EC1V 0HB, UK; M.Khader@city.ac.uk (M.A.K.); Martin.White@city.ac.uk (M.T.W.); A.Sayma@city.ac.uk (A.I.S.)
* Correspondence: salma.salah.2@city.ac.uk

Received: 28 May 2020 ; Accepted: 27 June 2020; Published: 23 July 2020

Abstract: Supercritical carbon dioxide (sCO_2) power cycles are promising candidates for concentrated-solar power and waste-heat recovery applications, having advantages of compact turbomachinery and high cycle efficiencies at heat-source temperature in the range of 400 to 800 °C. However, for distributed-scale systems (0.1–1.0 MW) the choice of turbomachinery type is unclear. Radial turbines are known to be an effective machine for micro-scale applications. Alternatively, feasible single-stage axial turbine designs could be achieved allowing for better heat transfer control and improved bearing life. Thus, the aim of this study is to investigate the design of a single-stage 100 kW sCO_2 axial turbine through the identification of optimal turbine design parameters from both mechanical and aerodynamic performance perspectives. For this purpose, a preliminary design tool has been developed and refined by accounting for passage losses using loss models that are widely used for the design of turbomachinery operating with fluids such as air or steam. The designs were assessed for a turbine that runs at inlet conditions of 923 K, 170 bar, expansion ratio of 3 and shaft speeds of 150k, 200k and 250k RPM respectively. It was found that feasible single-stage designs could be achieved if the turbine is designed with a high loading coefficient and low flow coefficient. Moreover, a turbine with the lowest degree of reaction, over a specified range from 0 to 0.5, was found to achieve the highest efficiency and highest inlet rotor angles.

Keywords: concentrated-solar power, supercritical carbon dioxide cycle, axial turbine design, micro-scale turbomachinery design.

---

1. Introduction

Micro-gas turbines coupled with concentrated-solar power systems (CSP) can provide a viable solution for renewable energy generation. They have been shown to be ideally suited for small-scale standalone and off-grid applications [1]. However, micro-gas turbines experience larger losses in the system components, and hence achieve lower thermal efficiencies, compared to large-scale gas turbines. Thus, for a high thermal efficiency, in the range of 40 to 50%, the system needs to operate at high heat-source temperatures, above 600 °C. In comparison, cycles operating with supercritical carbon dioxide (sCO_2) can achieve similar thermal efficiencies at more moderate temperatures. Therefore, sCO_2 can be considered as a potential candidate for concentrated-solar power applications, particularly for stand-alone solar dish units; offering a simple layout, high-power density and compact structures [2].

Despite the promising potential of sCO_2, sCO_2 turbomachine design is still a developing field. However, turbine performance is one of the main factors that affects the cycle performance; for example, a 2% increase in turbine efficiency has been shown to result in a 1% enhancement in the thermodynamic cycle efficiency [3]. This would have a significant impact on cost reduction of the solar power system through reduction in the size and the cost of the concentrator which typically represents over 60% of the total system cost. Therefore, several researchers have investigated advancing the state-of-the-art with regards to turbine design.

Moroz et al. [4] discussed some design aspects for a 100 MW axial sCO_2 turbine for a direct sCO_2 recompression cycle. Aerodynamic and structural analyses were performed to determine the best design configuration as a function of the number of stages, radial tip clearance androtational speed. Holaind et al. [5] addressed the design of small radial turbomachinery with an output power ranging from 50–85 kW and an efficiency of 70%. Subsequent to that, Qi et al. [6] presented a new insight for a sCO_2 radial turbine design for a power rating ranging between 100 and 200 kW through the integration of mean-line design with a loss model where an efficiency of 78–82% has been achieved. Saeed et al. [3] developed an algorithm that allows for the design of a sCO_2 radial turbine for CSP application rated at 10 MW. During the design process, a mean-line design tool that uses an enthalpy loss model, geometry optimisation and 3D RANS simulation were performed. As a result of the geometry optimisation, an enhancement in both the efficiency and power output of 5.34% and 5.30 % was achieved respectively [3]. In the same context, Zhou et al. [7] and Zhang et al. [8] proposed 300 kW and 1.5 MW radial sCO_2 turbine designs respectively. Zhang et al. [8] added a design of a 15 MW sCO_2 axial turbine to their study. They implemented CFD simulations to analyse the flow characteristics of sCO_2 turbine components. In the same year, Lv et al. [9] developed an optimisation design approach for a radial-inflow turbine using sCO_2. This was done through combining a one-dimensional design method with an optimisation algorithm for both nominal and off design performance conditions for the stage inlet temperature, rotational speed and expansion ratio. Likewise, Shi et al. [10] presented an optimal 10 MW three-stage sCO_2 axial turbine design using 3D model, optimisation methods and off design analysis. Considering the potential of sCO_2 fluid for small-scale systems, White et al. [11] presented a comparative study between various turbine architectures for a small-scale 100 kW sCO_2 Rankine cycle. Single-stage radial-inflow, single-stage axial, and two-stage axial turbines were analysed to identify the most feasible turbine designs based on the limitations of the blade height. It was found that the feasible blade height is in the range of 1.74 to 2.47 mm for a given turbine diameter of 30 mm. Furthermore, it was concluded that a low degree of reaction is preferred for single-stage turbines resulting in supersonic conditions at the rotor inlet. However, a higher degree of reaction is suitable for two-axial stage turbine leading to subsonic rotor inlet conditions.

Besides the aforementioned research work in the field of sCO_2 turbo-machines designs, a large amount of work has been conducted for micro-scale organic Rankine cycle (ORC) turbine design, using the same design methodology for novel working fluids, including refrigerants such as R134a, R1234yf and R152a. Fiaschi et al. [12] developed a design tool that included a loss model to examine the performance of a 50 kW radial turbine operating with various working fluids. The results of the study demonstrated that an efficiency ranging from 78–85 % can be achieved with a highest value (85%) for R134a and lowest value (78%) for R1234yf. Casati et al. [13] developed two preliminary designs for a five-stage transonic and eight stage supersonic radial-outflow 10 kW ORC turbine. It was found that the transonic turbines outperform the supersonic turbine during the partial-load operations thus resulting in enhanced efficiency. Rahbar et al. [14] proposed a mean-line model integrated with both an optimisation algorithm and a real-gas formulation for a 15 kW radial turbine. The design model resulted in a turbine efficiency ranging from 82.4–84% with a maximum value found for R152a. Lio et al. [15] integrated both mean-line design with loss correlations, developed previously for radial gas turbines, for examining turbine size, working conditions and predicting the efficiency of radial-inflow turbine operating with R245fa. The study revealed that the turbine size has a noticeable effect on the efficiency, and thus an efficiency of 85 to 90% at high expansion ratio has been attained at an output power greater than 50 kW.

Radial turbines are known to be an effective and compact machine for small-scale applications with a power ranging from 50 kW to 5 MW, allowing for the expansion of the working fluid in one single stage [12]. Thus, the radial-inflow turbine configuration has been the main candidate for small-size turbomachinery design in most of the aforementioned researches. Alternatively, a feasible single-stage axial turbine design could be achieved for micro-scale applications allowing for better heat transfer control, as the hot blades are far from the shaft, and subsequently prolongs the life of the bearing. In the current study, a mean-line approach is used to address the design of a single-stage 100 kW sCO$_2$ axial turbine design through identifying optimal turbine design parameters from both mechanical perspectives, owing to the significant axial thrust loads, high rotational speeds and high operating pressures, and aerodynamic performance perspectives. Additionally, the design tool implemented in the current study is refined by introducing Soderberg's and Ainley and Mathieson's loss correlations, which allows the loss within the rotor and stator to be estimated based on loading coefficient, tip clearance and blade geometry. The designs were evaluated at three different speeds of 150k, 200k, 250k RPM respectively,to achieve an overall turbine efficiency greater than 80% and keep low centrifugal stresses on the rotor blades, for turbine inlet conditions of 923 K, 170 bar, expansion ratio of 3. For the given inlet conditions, a single-stage design is proposed due to the low power rating and low volume ratio of the machine. The novelty in this current work lies in presenting a design of sCO$_2$ micro-scale single-stage axial turbine, alongside with defining the optimal turbine design parameters from both mechanical and aerodynamic performance perspectives.

This paper is structured as follows: an overview of turbine design methodology and assumptions are presented in Section 2. Axial turbine design and loss modelling is discussed in Section 3. The results and discussion are presented in Section 4, before the final conclusions are presented.

2. Turbine Design Methodology

sCO$_2$ condensation (transcritical) power cycles were firstly proposed by Angelino and Feher in 1968 where efficiencies up to 50 % can be achieved [16,17].However, to achieve condensation within the cycle it is necessary to lower the temperature of the CO$_2$ below its critical temperature (31.1 °C). Unfortunately, achieving this within CSP plants, which are typically in dry arid regions with high solar irradiation, requires water-cooling which is usually not feasible in these sites. To overcome this issue, it has been proposed to raise the critical point of the working fluid by doping CO$_2$ with another fluid. For example, mixing CO$_2$ with various additives, such as C$_3$H$_8$, C$_4$H$_8$, C$_4$H$_{10}$, C$_5$H$_{10}$, C$_5$H$_{12}$ and C$_6$H$_6$, has been shown to increase the critical temperature to up to 60 °C [18]. Moreover, the application of sCO$_2$ blends in CSP plants have been previously studied [18] and have been shown to enhance efficiency of a Brayton power cycle by 3–4% compared to pure sCO$_2$. Therefore, sCO$_2$ blends have been proposed instead of pure sCO$_2$ to increase the critical temperature of the working fluid [19,20]. The presented design methodology has been developed as part of a preliminary study related to the Horizon 2020 SCARABEUS project [21], and thus the methodology is capable of designing turbines intended for sCO$_2$ blends. However, for simplicity at this stage, pure sCO$_2$ will be considered as the working fluid. Considering a blend merely changes the inputs into the equation of state used to predict thermodynamic properties and thus the method can be readily extended to blends without any change to the analysis. It should be noted that the turbine design is likely to be sensitive to the chosen blend, but such investigations are left for a future work.

The turbine design is part of a condensation sCO$_2$ cycle, as shown in Figures 1 and 2, with a net power output of 100 kW. The turbine inlet parameters are obtained from a thermodynamic cycle analysis and are selected as a compromise between cycle efficiency, component life, complexity and feasibility [11]. Accordingly, the turbine inlet temperature is limited to 650 °C. The turbine inlet pressure is limited to 17 MPa with an expansion ratio of 3 to reduce the power block weight, size and price [22]. The cycle is designed assuming a pump inlet temperature of 20 °C, pressure of 6 MPa, compressor isentropic efficiency of 70%, turbine isentropic efficiency of 80% and recuperator effectiveness of 90%. For the selected parameters, the cycle has a thermal efficiency of 32.8% with a mass-flow rate of 0.65 kg/s.

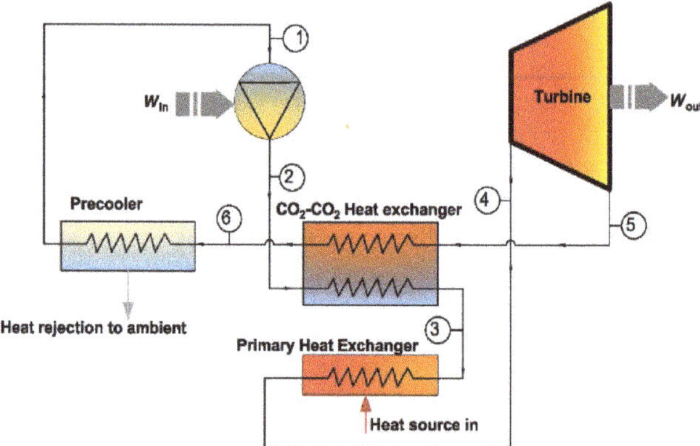

Figure 1. Cycle schematic drawing.

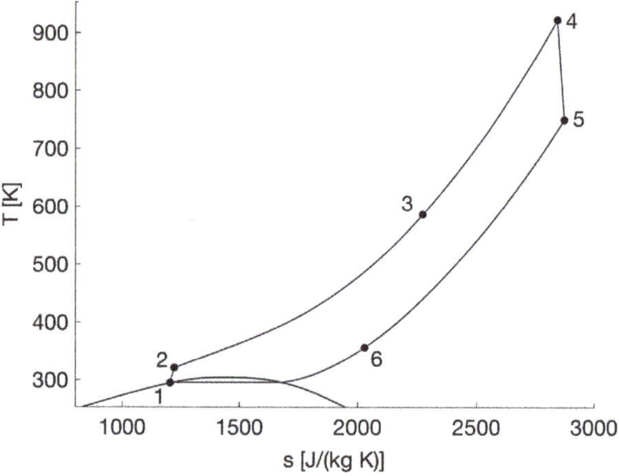

Figure 2. Entropy versus temperature (*T-s* diagram) for the thermodynamic cycle in Figure 1.

To design the turbine, a mean-line turbine design approach is adopted to provide a fast and accurate estimation of the turbine geometry and the expected isentropic efficiency. This has been developed in MATLAB and coupled with NIST REFPROP [23] to account for the sCO$_2$ fluid behaviour. The model is also capable of operating with Simulis, which enables novel fluid blends to be considered [24]. Within the model the steady-state mass, energy and momentum equations are solved to obtain the geometric parameters of the turbine. In this section, the turbine mean-line design is developed along with a parametric study to examine the effect of changing various design parameters on the turbine performance, at various rotational speeds, and assess blades pressure and centrifugal loads. The designs were evaluated for turbine inlet conditions of 923 K, 170 bar, expansion ratio of 3.

For brevity, a full description of the model is not presented here, but can instead be found in Appendix A. Instead, a brief overview of the process is provided here. The process starts with the choice of the non-dimensional parameters, namely the flow coefficient (ϕ), and loading coefficient (ψ), which are defined as:

$$\phi = C_a/U, \tag{1}$$

$$\psi = \frac{2\Delta h_{os}}{U^2}, \tag{2}$$

where Δh_{os} is the enthalpy drop across the stage, U is the blade velocity and C_a is the axial velocity. Alongside these, the degree of reaction Λ and an initial estimate for the total-to-total isentropic efficiency are also specified. The values of the mass flow rate, turbine inlet temperature and pressure are taken from the cycle analysis. The selection of the dimensionless parameters is made with the aim of maximising turbine efficiency; the first two parameters, ϕ and ψ, according to typical values taken from the Smith chart [25], while the degree of reaction is assumed to be 0.5 [26]. Then, the loss coefficients are calculated using the selected loss model; where they are chosen to obtain an accurate estimation for the profile, secondary and tip clearance losses at the design operating conditions as detailed in Section 3. Following this, the design is re-iterated using the calculated loss coefficients and the estimate for the turbine efficiency is updated. Once the design geometry is obtained, the blade heights and mean diameter are calculated as a function of the area, mean velocity and rotational speed, and hence the feasibility of the design can be verified in comparison to the feasible manufacturing values for the inlet blade height and blade diameter [11].

Furthermore, pressure and centrifugal loads are examined where Ni-Cr-Co alloy (Inconel 718) is assumed for the turbine blades; it has been considered as a suitable material in the past [26]. The loads are calculated, assuming tapered blades with an average density of 8000 kg/m^3 and a maximum equivalent stress of 303 MPa, using Equations (A19) and (A20) [26,27]. The centrifugal load is calculated at the mean blade height and mean rotor area. To calculate the bending load, the number of rotor blades are defined as function of the blade pitch (s) at the mean radius (r_m):

$$n_R = 2\pi r_m/s, \tag{3}$$

The optimum pitch to chord ratio is obtained as a function of the flow angles and the chord length is obtained assuming an aspect ratio of 1 [28]. The design steps and methodology are summarised in Appendix A and in the flow chart (Figure 3).

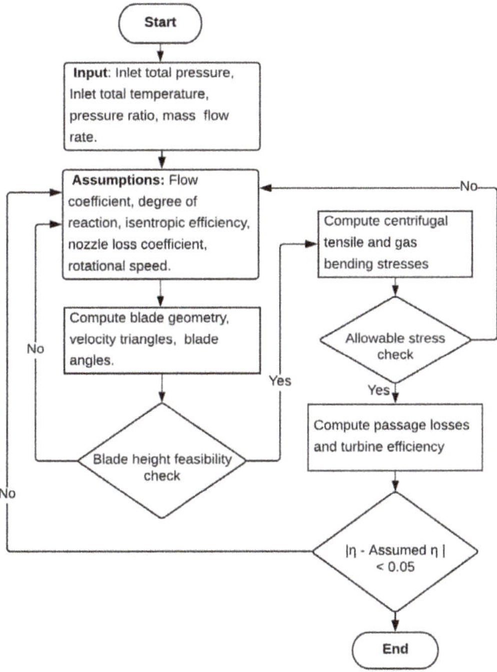

Figure 3. Design flow chart.

In principle, the optimal axial turbine efficiency occurs within a certain range of specific speed, typically $0.1 < N_s < 1.0$ [29]. For the specified specific speed range, the rotational speed ranges varies between 64 and 640 kRPM. Running the turbine at higher specific speed imposes high level of stresses on the rotor blades and could also result in rotordynamic instabilities. Consequently, in the present study the turbine design will be evaluated at three different shaft speeds, namely 150k, 200k and 250k RPM, corresponding to specific speed ranging from 0.23 to 0.39 rad, to achieve an overall turbine efficiency greater than 80%. The shaft speed is determined using Equation (4), by assuming that the specific speed is within the optimal range for axial turbines:

$$N_s = \frac{\omega \sqrt{\dot{Q}}}{\Delta h_{os}^{3/4}},\tag{4}$$

In Equation (4), ω is the rotor rotational speed in rad/s, $\dot{Q}$ is the volumetric flow rate at the rotor outlet in m^3/s and Δh_{os} is the isentropic enthalpy drop across the turbine, in J/kg.

To examine the effect of the various design parameters on the turbine performance and the feasibility of the design, the flow coefficient and loading coefficients have been varied over the range of 0.2 to 1 and 0.8 and 3 respectively [25]. Also, the the degree of reaction is varied from from 0–0.5 [26]. Then, the radius for root, mean, tip profiles are obtained using the free vortex design equations. Table 1 reports the range of values selected for the various design parameters.

Table 1. Input design specifications.

Design Parameter	Value	Design Parameter	Value
Turbine inlet temperature	650°C [11]	Turbine inlet pressure [MPa]	17 [22]
Expansion ratio [-]	3.0	Net power output [kW]	100
Rotational Speed [kRPM]	150–250	Flow coefficient [-]	0.2–1.0 [25]
Degree of Reaction [-]	0.0–0.5 [26]	Loading Coefficient [-]	0.8–3.0 [25]
Ni-Cr-Co density [kg/m³]	8000	Inconel 718 equivalent stress [MPa]	303 at 1073 K [27]

3. Loss Modelling

To predict axial turbine performance, various loss models have been previously introduced starting from Soderberg [30], and then Ainely and Mathieson [31]. This has been followed by modifications presented by Dunham and Came [32], Craig and Cox [33], Kacker and Okapuu [34], and finally off-design correlations proposed by Moustapha et al. [35]. Soderberg's loss model accounts for the effect of profile and secondary flow losses, while tip clearance and trailing edge losses are ignored. Profile losses are calculated as a function of the flow deflection while the secondary losses are interpreted as function of the aspect ratio neglecting the effect of inlet boundary layer and blade geometry. While Soderberg's model is considered to be an oversimplified model where the effect of Mach number (Ma) and fluid non-dimensional parameters are neglected, it is considered to be satisfactory for preliminary design phase as it allows the loss within the stator and rotor to be estimated based on the amount of expansion that occurs within each passage [36].In the presented design framework, velocities approach the sonic speed and therefore Soderberg model is considered to be more accurate for estimating the flow losses as the correlations were derived based on high Mach number data. In the absence of a tip clearance loss correlation in this model, the Ainley and Mathieson correlation has been used. It is worth mentioning that a comparative study was made between using Ainley and Mathieson correlations only and the combination of the two classes, and the same trends were observed with a deviation in the design point efficiency of approximately 1.65%. Equations (5)–(7) represent the loss coefficients predicted using Soderberg model and Equations (10)–(11) represent the tip clearance loss coefficient estimated using the Ainely and Mathieson model.

In Soderberg's model, the losses are modelled as a function of the aspect ratio (H/l), the nominal loss coefficient ε^* and the blade deflection ε. The nominal loss coefficient is obtained from the empirical correlation:

$$\zeta^* = 0.04 + 0.06\left(\frac{\varepsilon}{100}\right)^2,\tag{5}$$

where ε is the flow deflection angle, which corresponds to $\varepsilon_N = \alpha_1 + \alpha_2$ for the stator and $\varepsilon_R = \beta_2 + \beta_3$ for the rotor. The nominal loss coefficients are then obtained as follows:

$$\zeta_N = \left(\frac{10^5}{Re}\right)^{1/4}\left[(1+\zeta^*)\left(0.993 + 0.075\frac{l}{H}\right) - 1\right],\tag{6}$$

$$\zeta_R = \left(\frac{10^5}{Re}\right)^{1/4}\left[(1+\zeta^*)\left(0.975 + 0.075\frac{l}{H}\right) - 1\right],\tag{7}$$

In Ainely and Mathieson's model, the tip clearance coefficient Y_k is defined in terms of pressure drop as a function of the pitch to chord ratio (s/c), average blade angle β_m, the radial tip clearance (k), the average blade height (h) and a constant (B).

$$\beta_m = \tan^{-1}[(\tan\beta_3 - \tan\beta_2/2)],\tag{8}$$

$$C_L = 2\ (s/c)\ (\tan\beta_2 + \tan\beta_3)\cos\beta_m,\tag{9}$$

where B equals to 0.5 for radial tip clearance. Considering the manufacturing tolerances and uncertainties in thermal expansion during the operation of micro-scale turbines, the radial tip clearance for the un-shrouded blades is set to 0.1 mm [6]. The tip clearance loss coefficient is then obtained as follows:

$$Y_k = B\ \left(\frac{k}{h}\right)\left[\frac{C_L}{s/c}\right]^2\left[\frac{\cos^2\beta_3}{\cos^3\beta_m}\right],\tag{10}$$

The enthalpy loss coefficient for the rotor (λ) is then obtained as a function the rotor blade exit actual relative and isentropic temperatures (T_{03rel}) and (T_3'') respectively, the rotor exit velocity (V_3) and the specific heat capacity of the working fluid (c_p).

$$\lambda_k = \frac{Y_k}{(T_{03,rel}/T_3'')}\tag{11}$$

4. Model Validation

One of the challenges that face sCO$_2$ turbine design is the lack of experience and published validation data available within the literature. Most of the experimental test rigs available for sCO$_2$ turbomachines have considered small-scale radial turbines (i.e. Sandia national laboratory (SNL), Naval nuclear laboratory (NNL), The Tokyo institute of technology (TIT), Korea atomic energy research institute (KAERI) among others) [37–40]. Furthermore, previous sCO$_2$ turbine designs mostly focus on large-scale turbines, for which the axial turbine is the preferred configuration. To support this study, the design model has been verified against a study conducted for a 100 MW sCO$_2$ axial turbine design presented by Schmitt et al. [41]. In their study, a first-row aerodynamic analysis for a six-stage 100 MW sCO$_2$ Brayton cycle turbine was presented. The mean-line design, integrated with Soderberg's correlation, was implemented to predict aerodynamic losses. The results of the mean-line design were verified with a 3D simulation using STAR-CCM+; which verified the 1D mean-line design. Using the same design inputs from that study, the model developed in this work, ignoring the tip clearance losses, has been used to design the same turbine and good agreement between the two models is found (Table 2) [41]. Whilst the presented validation provides a preliminary validation for the design model, further studies (i.e., CFD and experimental) are required to ensure the suitability of the implemented loss models for sCO$_2$ micro-scale axial turbines; these models were derived for air turbines and have not been validated yet for sCO$_2$ turbines.

Table 2. 100 kW supercritical carbon dioxide (sCO$_2$) turbine design model validation.

Performance Parameters	Verified [41]	Results	Difference [%]	Design Angles	Verified [41]	Results	Difference [%]
η_{tt}	0.903	0.9058	0.31	β_3	42.3	42.310	0.024
η_{ts}	0.835	0.8391	0.48	β_2	63.9	63.960	0.094
ζ_R	0.108	0.1077	0.28	α_2	70.7	70.703	0.004
ζ_N	0.070	0.0699	0.14	α_3	5.70	5.777	1.340

5. Results and Discussion

A parametric study is presented in this section to investigate the effect of the flow coefficient (ϕ), degree of reaction (Λ) and loading coefficient (ψ) on the turbine performance η_{tt} and design feasibility. Different turbine designs are generated assuming different values of the design parameters ϕ, ψ, Λ, based on the specified ranges in Table 1. Accordingly, a Smith chart is obtained as shown in Figure 4, which shows the normalised efficiency achieved for the sCO$_2$ turbine with a degree of reaction and rotational speed of 0.5 and 150 kRPM respectively.

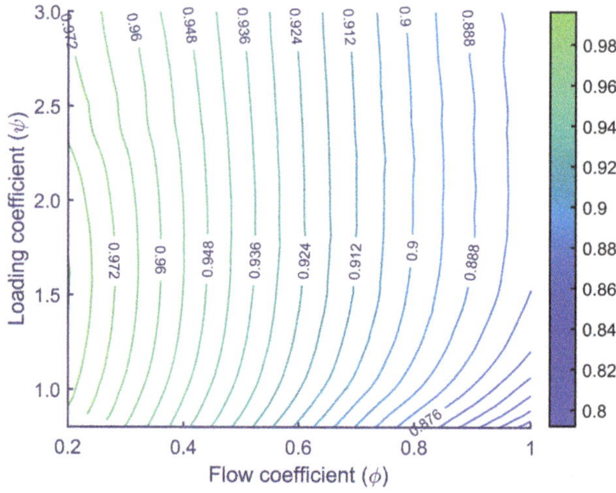

Figure 4. Contour plot for normalised sCO$_2$ turbine efficiency at $\Lambda = 0.5$ and 150 kRPM.

It is observed that the highest normalised efficiencies are obtained at low flow and loading coefficients, as in the original Smith chart [25]. According to Figure 4, a design point can be selected as a starting point for the parametric study. Though one of the benefits of using sCO$_2$ is having compact component designs, clearance losses will be proportionally larger compared to turbomachines of a comparable power rating owing to the high density of sCO$_2$ and small turbine dimensions. Additionally, windage losses could be expected to be significant on the turbomachinery wheel surfaces compared to large-scale gas turbines as reported from the tests conducted by the Naval Nuclear Laboratory and the Tokyo institute of technology's (TIT) [39,40]. Therefore, whilst windage losses are not accounted for in the current study, it is important to include these in the future. Likewise, losses due to surface roughness are not considered in this analysis owing to the simplicity of the implemented loss model. Thus, the mean-line model will be extended to include both roughness and windage effects in future studies.

The effect of changing the rotor rotational speed on η_{tt} at different flow and loading coefficients has also been examined. The efficiency is found to increase linearly with increasing rotational speeds over the range from 150 to 250 kRPM for the loading and flow coefficient domains specified in Table 1. The lowest flow coefficient, over the range from 0.2 to 1.0, results in the highest efficiency level, as confirmed in Figure 4. However, a low flow coefficient results in larger turbine annulus area and a larger deflection angle. A loading coefficient of 1.6 results in the maximum efficiency at a fixed flow coefficient and degree of reaction of 0.2 and 0.5 respectively. At a loading coefficient of 1.6, the swirl angle is kept close to the recommended value, which is recommended not to exceed $20°$ [26], whilst the optimum difference in the whirl velocity components at the inlet and exit of the rotor is achieved, which enables a high efficiency to be achieve. Thus, for this design, the flow coefficient and loading coefficient are set to 0.2 and 1.6 respectively to achieve the maximum efficiency at a degree of reaction of 0.5. The corresponding turbine power output is 116 kW, with a total-to-total isentropic efficiency of 78%.

In the following set of results, Figure 5a–d, ϕ and Λ have been set to 0.2 and 0.5 respectively, whilst ψ is varied from 0.8 to 3.0. Increasing the loading coefficient from 0.8 to 3.0 results in an increase in both rotor absolute and relative inlet flow angles, α_2 & β_2, from 74 to $81°$ and -60 to 60 $°$ respectively as shown in Figure 5a. In principle, low loading coefficients result in higher efficiency, though it results in higher blade speed and thus high mechanical stresses; at values of 1.0 and 3.0 the mean blade speed (U_m) reaches approximately 522 and 306 m/s, and results in a total blade stresses of 160 MPa and 76 MPa respectively. Meanwhile, at high loading values a smaller number of rotor blades is needed. Therefore, the design decision should be made based on the selected material, the maximum allowable stress, along with the required number of blades.

Furthermore, increasing the loading coefficient causes a slight decrease in both Ma_2 and Ma_3, at the inlet and outlet of the rotor blades respectively, as shown in Figure 5b; where Ma_2 and Ma_3 are the Mach numbers calculated with respect to the absolute and the relative velocity respectively. To achieve a subsonic flow at a flow coefficient of 0.2, the loading coefficient should be greater than 0.8.

Additionally, it was found that increasing the loading coefficient results in an efficiency increase until a maximum is reached at values of ψ ranging between 1.6 to 1.7 for the three rotational speeds. It's worth mentioning the ψ has a limited effect on the efficiency at all rotational speeds for a flow coefficient of 0.2. The highest efficiency is achieved for the turbine designs that keep the swirl angle α_3 close to the recommended value, which is recommended not to exceed $20°$ [26], along with achieving the optimum difference in the whirl velocity components at the inlet and exit of the rotor. Increasing the exit circumferential velocity, owing to the increased swirl angle, results in an increase in centrifugal force which leads to an increased amount of flow reversal at the rotor outlet. Hence, higher losses and lower efficiencies are observed.

Increasing ψ results in an increase in the blade height at the rotor inlet and outlet blade heights reaching, up to 2.1 and 3.7 mm for b_2 and b_3 respectively at a speed of 250 kRPM (Figure 5d). For feasible wheel manufacturing, it is hypothesised that inlet blade height and blade diameter should be designed to be above a minimum allowable rotor diameter of 30.00 mm and a minimum allowable blade height b_2 of 1.25 mm [11]. Thus, according to these criteria, for a design with a flow coefficient of 0.2 and rotational speed of 250 kRPM, ψ should be above 1.7. Owing to the micro-scale design dimensions, it is anticipated that micro-electrical discharge machining (EDM) milling could be used for the manufacturing of the turbine components. EDM is already in use for micro-gas turbines where high accuracy features, in the order of several micrometres, and surface finishes, with roughness values as low as 0.4 μm, can be achieved [42]. Similar to this design scale, Korea institute of energy research (KIER) developed a prototype for sCO_2 Brayton cycle in which a 60 kW axial impulse turbine was manufactured; this demonstrates the ability to manufacture the turbine and the applicability of an axial turbine configuration for micro-scale applications [43].

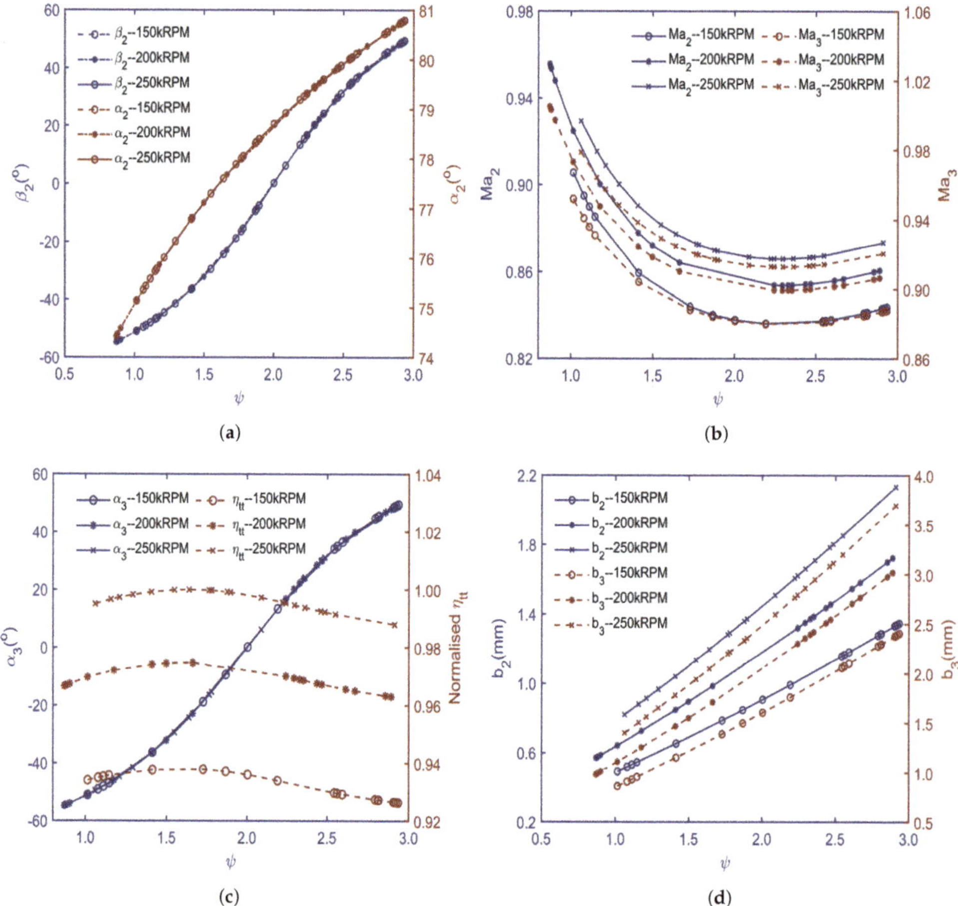

Figure 5. Loading coefficient (ψ) versus (**a**) flow angles [β_2 and α_2] (**b**) Mach number at the rotor inlet [Ma_2] and exit [Ma_3] (**c**) normalised efficiency [η_{tt}] and swirl angle [α_3] (**d**) blade heights [b_2 and b_3] at various rotational speeds.

To investigate the effect of changing the degree of reaction on the performance of the axial turbine, Λ has been varied between 0.0 and 0.5 while fixing ϕ to 0.2 and ψ to 1.6. Accordingly, the flow angle β_2 decreased from 61 to $-26°$ and α_2 decreased from 82 to 78° as shown in Figure 6a. In the same context, increasing the degree of reaction results in an increase in Ma_3 from 0.50 to 0.92 and a decrease in Ma_2 from 1.43 to 0.90 as shown in Figure 6b. At low reaction values, the stator outlet velocity is high as a result of the large acceleration and thus the Mach number is expected to be high. A higher degree of reaction results in a thin boundary layer and less tendency to secondary flow as result of having a good acceleration at the stator outlet [44].

A noticeable decrease in the efficiency has been experienced while increasing the degree of reaction from 0.0 to 0.5, and this is found for all three of the rotational speeds considered. Particularly, the normalised efficiency decreased from 0.97 to 0.91 at a rotational speed of 150 kRPM. At high degree of reaction values, the swirl angle is small which results in less rotor losses. However, high degree of reaction results in low rotor inlet blade angles (α_2), and hence an overall reduction in the efficiency as observed in Figure 6c. In view of the fact that high reaction leads to higher pressure and high density at the rotor inlet, the blade heights decrease as the degree of reaction is increased, as shown in Figure 6d.

A maximum value of 2.20 mm was found at zero reaction and 250 kRPM and a minimum of 0.74 mm at 150 KRPM and reaction of 0.5.

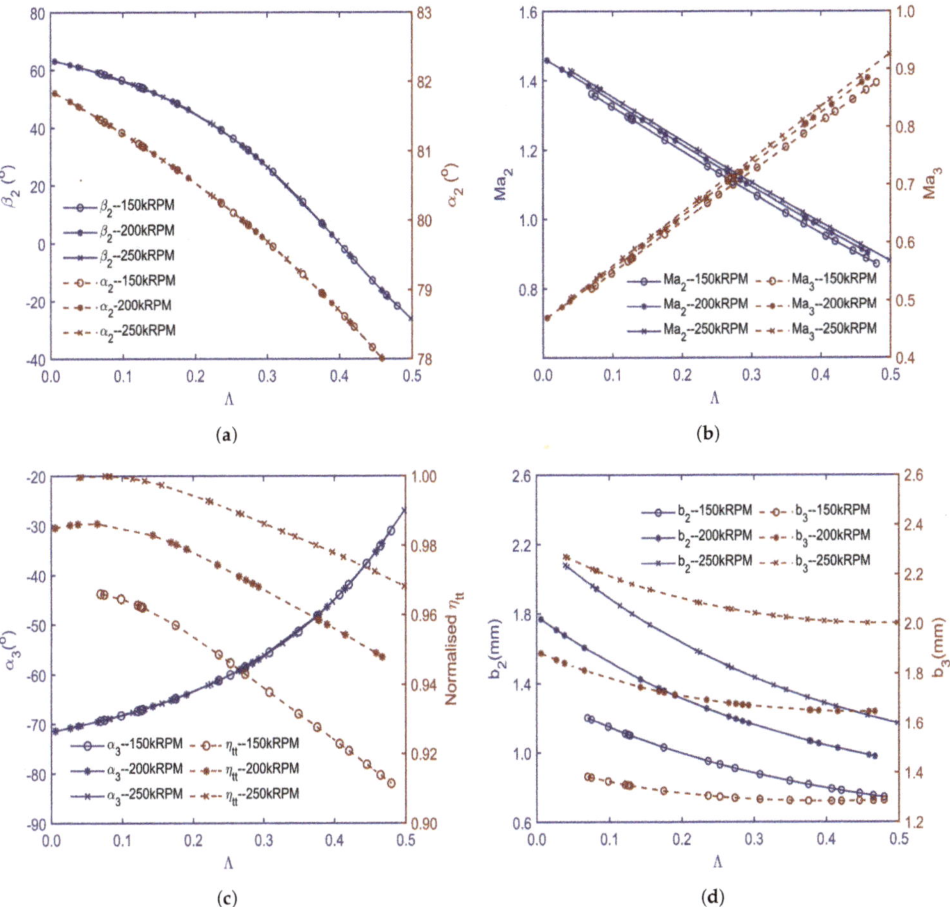

Figure 6. Degree of reaction versus (Λ) (**a**) flow angles [β_2 and α_2] (**b**) Mach number at the rotor inlet [Ma_2] and exit [Ma_3] (**c**) normalised efficiency [η_{tt}] and swirl angle [α_3] (**d**) blade heights [b_2 and b_3] at various rotational speeds.

To investigate the effect of the flow coefficient on the turbine performance, the analysis has been repeated at $\Lambda = 0.5$ and $\psi = 1.60$ respectively. Reducing ϕ over the range from 0.2 to 1 results in an increase in rotor outlet flow angle α_2 from 43 to 77° as shown in Figure 7a. However, increasing the flow coefficient also resulted in an increase in both Mach numbers, as shown in Figure 7b, where Ma_2 increased from 0.88 to 1.30 and Ma_3 increased from 0.93 to 1.34 at 250 kRPM; thus, supersonic conditions occur in both the rotor and stator. To ensure subsonic flow at the rotor inlet, the flow coefficient should be kept below 0.45 for a loading coefficient of 1.6 and degree of reaction of 0.5.

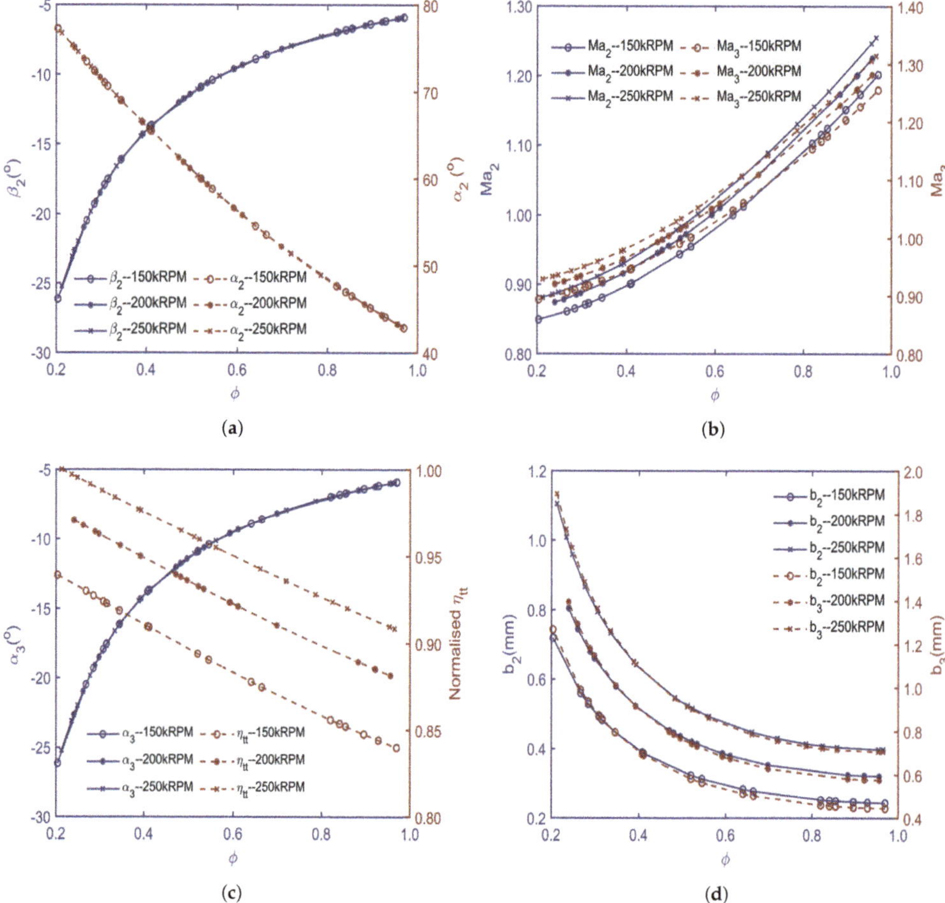

Figure 7. Flow coefficient (ϕ) versus (**a**) flow angles [β_2 and α_2] (**b**) Mach number at the rotor inlet [Ma_2] and exit [Ma_3] (**c**) normalised efficiency [η_{tt}] and swirl angle [α_3] (**d**) blade heights [b_2 and b_3] at various rotational speeds.

Furthermore, increasing the flow coefficient results in an efficiency decrease at all rotational speeds, as observed in Figure 7c. Additionally, it results in a decrease in swirl angle from 26 to 5°. However, at high flow coefficients, low swirl angles and low rotor inlet blade angles (α_2) are achieved. Increasing the flow coefficient results in higher Mach number at the rotor exit and hence higher losses incurred by the formation of shock waves in the rotor blade passages. Additionally, profile losses will be higher owing to the boundary layer growth, whilst friction losses at the exit are expected to be high. Consequently, a drop-in efficiency is observed at higher flow coefficients. The effect of changing the flow coefficient on the blade heights is shown in Figure 7d. Increasing the flow coefficient resulted in a decrease in both blade heights at all speeds. A minimum inlet blade height of 0.24 mm is found at a rotational speed of 150 kRPM and flow coefficient of 1.0.

The effect of changing the aspect ratio on the turbine efficiency is shown in Figure 8a. Increasing the aspect ratio from 1 to 3 resulted in the normalised efficiency increasing from 0.987 to 1.000 at 250 kRPM. Higher aspect ratios result in lower rotor and stator losses and higher efficiency. For high aspect ratios, secondary effects are confined to the end-wall region. However, it affects the whole passage for small aspect ratios [44]. The gas bending and centrifugal tensile stresses have been evaluated for the selected material at a design condition of $\phi = 0.2$, $\psi = 1.6$, $\Lambda = 0.5$, 150 kRPM and aspect ratio of 1.At this point, the gas bending and centrifugal tensile stresses are predicted to be 122 and 35 MPa respectively; resulting into a total stress of 157 MPa, which falls within the material allowable limit ($\sigma_{max} = 303$ MPa). A creep failure criterion, which shows the continuous application of a steady stress over a period of time at various temperatures required to produce 0.2 percent strain, is used to assess blade life. Based on the results from [26], it is anticipated that a blade life of 10,000 hr can be achieved under these operating conditions. It is worth emphasising that creep failure is an important consideration, but a detailed analysis is outside the scope of this study.

A noticeable difference between the magnitude of the gas bending (GB) stress in comparison to the centrifugal tensile (CT) stress, for tapered blades, is experienced owing to the density effect of sCO$_2$. Increasing the aspect ratio from 1 to 3 results in an increase in the gas bending stress by a factor of approximately 9. Furthermore, decreasing the degree of reaction from 0.5 to zero results in a decrease in the the gas bending stress by a factor of approximately 4. Hence, the aspect ratio and the degree of reaction should be minimised. However, decreasing the aspect ratio is associated with a drop in efficiency. Therefore, selecting the optimal aspect ratio is a trade-off between high efficiency and low stresses. Increasing the rotational speed from 150 to 250 kRPM results in centrifugal tensile stress increasing from 35 to 95 MPa as shown in Figure 8b.

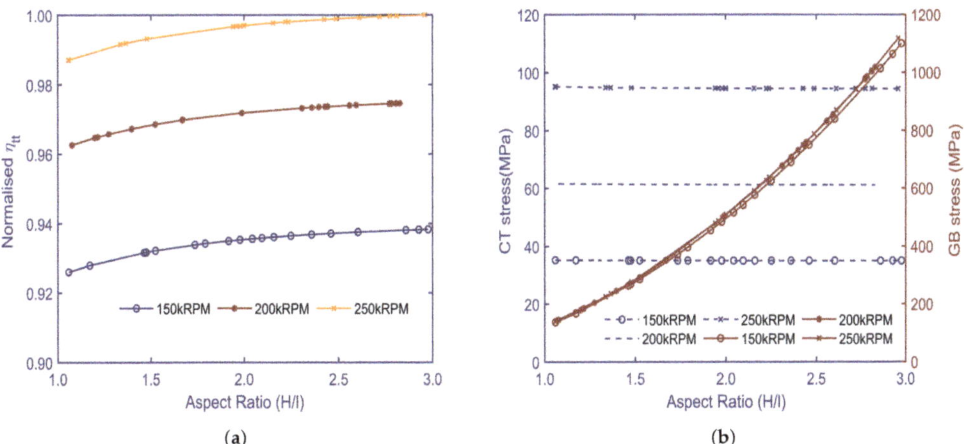

Figure 8. Aspect ratio of the rotor blades versus (**a**) normalised efficiency [η_{tt}] (**b**) centrifugal tensile [CT] and gas bending [GB] stresses at various rotational speeds.

The last part of the parametric study aims to show how the turbine design parameters can be varied to obtain a feasible turbine geometry, defined by a minimum allowable rotor diameter and inlet blade height of 30.00 mm and 1.25 mm respectively, over the range of the rotational speeds from 150 to 250 kRPM. The feasibility criteria can be achieved by operating at a high loading coefficient, where the diameter and inlet blade height are largest. Increasing the rotational speed results in a decrease in the diameter and an increase in the blade height. Hence, this should be considered during the selection of the rotational speed and flow coefficient of the turbine as shown in Figure 9a,b. The results reported in Figure 9a refer to 0.5 degree of reaction, while those in Figure 9b refer to zero degree of reaction. Thus, it can be concluded that running at a zero reaction (impulse) results in a more feasible range of inlet blade heights and diameters. The maximum diameter that can be achieved is 56 mm with a maximum inlet blade height of 3.3 mm at a flow coefficient of 0.2.

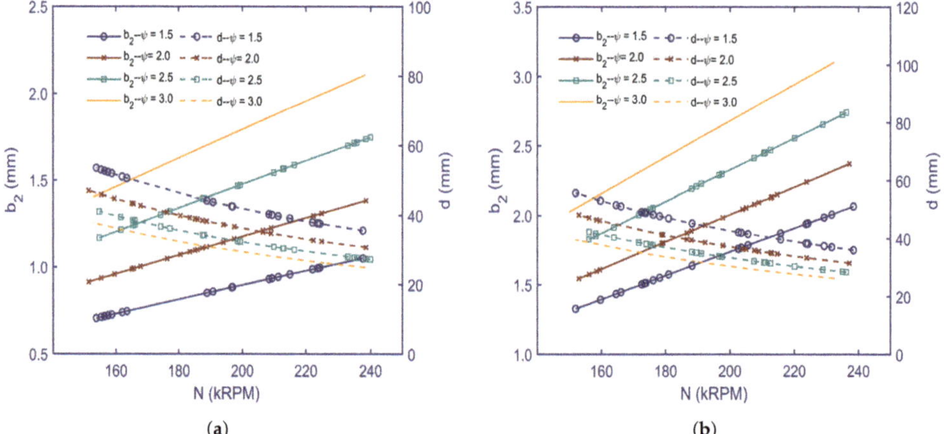

Figure 9. Rotor rotational speed versus the inlet blade height and mean diameter at (**a**) 0.5 degree of reaction (**b**) zero degree of reaction; blade height [b_2] is represented by the solid lines and diameter [d] is represented by the dashed lines.

A comparison between two turbine designs, to depict the effect of the selected design variables on the final turbine geometry, is presented in Figure 10. Geometry (A) is obtained at $\Lambda = 0.5$, $\psi = 1.6$ and $\phi = 0.2$, while geometry (B) is obtained at $\Lambda = 0.0$, $\psi = 3.0$ and $\phi = 0.2$ at 150 kRPM. Reducing the degree of reaction and increasing the loading coefficient results in more feasible design dimensions (Design B).

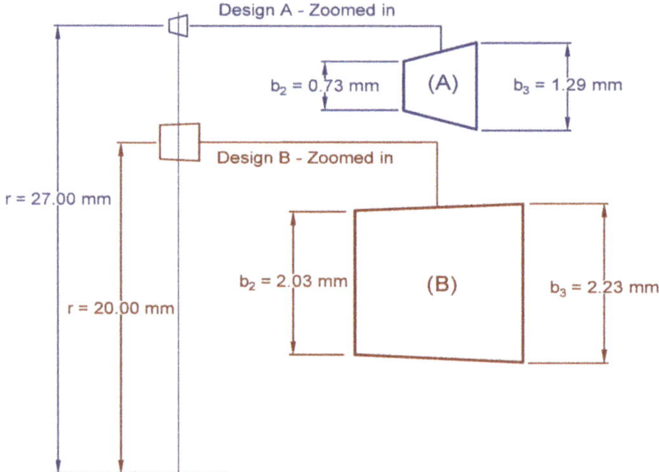

Figure 10. Comparison between two turbine designs at: (**A**) $\psi = 1.6$, $\phi = 0.2$ and $\Lambda = 0.5$; and (**B**) $\psi = 3.0$, $\phi = 0.2$ and $\Lambda = 0.0$.

6. Conclusions

This paper has presented the design of a small-scale single-stage 100 kW axial turbine for implementation within a supercritical carbon dioxide power system. The performance of the turbine was evaluated by introducing Soderberg's and Ainley and Mathieson's loss correlations in the design model. To verify the developed model, it was cross checked with published results for a larger turbine design which gave confidence in the methodology. A parametric study was conducted to investigate the effects of the degree of reaction, flow coefficient and loading coefficient on the performance and the feasibility of the proposed design. The designs were assessed at three rotational speeds, namely 150, 200 and 250 kRPM, with specific speed ranging between 0.23 to 0.39 rad. It is found that feasible designs could be obtained, considering the minimum allowable inlet blade height and mean diameter, at a low flow coefficient (in the range of 0.2), a high loading coefficient (>2.8), and a low degree of reaction at a rotational speed of 150 kRPM. Additionally, low degree of reaction results in higher efficiencies owing to the high rotor inlet blade angles. High aspect ratios were found to result in high efficiencies and high stresses, and hence the selection of the optimum aspect ratio is a trade-off between the high efficiency and low stresses.

This work has demonstrated the suitability of Soderberg's and Ainley and Mathieson's models loss correlations to investigate the design of a single-stage axial turbine for sCO_2 applications. It has also provided important insights into the trade-offs between aerodynamic and mechanical design that should be considered at this scale. However, the next necessary steps would be to conduct a more detailed structural analysis, alongside extending the mean-line model to include more sophisticated loss models, and conducting 3D CFD simulations to further validate the model.

Author Contributions: All the named authors designed the presented study, A.I.S and M.T.W. provided the paper conceptualization. The Methodology was conducted by S.I.S. The software was developed by S.I.S and M.T.W. S.I.S conducted the validation, formal analysis and investigation. S.I.S and M.A.K prepared the initial manuscript and M.T.W and A.I.S reviewed and edited the paper before submission. This work has been completed under the supervision of A.I.S. and M.T.W. All authors have read and agreed to the published version of the manuscript.

Funding: The SCARABEUS project has received funding from the European Union's Horizon 2020 research and innovation programme under grant agreement No. 814985.

Abbreviations

The following abbreviations are used in this manuscript:

A	Mean blade area [m^2]
b_2	Rotor inlet blade height [mm]
b_3	Rotor exit blade height [mm]
CSP	Concentrated-solar power
CT stress	Centrifugal tensile stress [MPa]
c	Blade chord length [mm]
C	Absolute Velocity [m/s]
C_a	Axial velocity component
d	Mean blade diameter [mm]
GB stress	Gas bending stress [MPa]
h	Mean blade height [mm]
KIER	Korea institute of energy research
KAERI	Korea atomic energy research institute
$\dot{m}$	Mass flow rate [kg/s]
Ma	Mach number [-]
EDM	micro-electrical discharge machining
n_R	The number of rotor blades [-]
N_s	Specific speed [rad]
N	Rotational speed [kRPM]
NNL	Naval Nuclear Laboratory
$\dot{Q}$	Volume flow rate [m^3/s]
r_m	Mean blade radius [mm]
RANS	Reynolds-averaged Navier–Stokes equations
RPM	Revolutions per minutes
sCO_2	Supercritical carbon dioxide.
SNL	Sandia national laboratory

TIT	The Tokyo institute of technology
U_m	Rotor blade mean velocity [m/s]
U	Blade velocity [m/s]
V	Relative Velocity [m/s]
s	Blade pitch [mm]
α	Absolute flow angle [°]
β	Relative flow angle [°]
Δh_{os}	Enthalpy drop across the stage [kJ/kg]
Δh_o	Enthalpy drop across the entire turbine [kJ/kg]
ζ	Enthalpy loss coefficient [-]
η_{tt}	Total to total Efficiency [%]
η_{ts}	Total to static Efficiency [%]
λ	Enthalpy loss coefficient [-]
Λ	Degree of reaction [-]
ρ_b	Density of blade material [kg/m^3]
ψ	Blade loading coefficient [-]
ϕ	Flow coefficient [-]

Appendix A. Design Steps

From the thermodynamic specification of the turbine and an assumed turbine efficiency the enthalpy drop across the entire turbine Δh_o can be readily found. The blade speed U and axial velocity component C_a then follow from the loading coefficient ψ and flow coefficient ϕ:

$$U = \sqrt{\frac{2\Delta h_{os}}{\psi}} \tag{A1}$$

$$C_a = U\phi \tag{A2}$$

The blade angles for the stator and rotor are determined as a function of the blade speed U, the loading coefficient ψ, flow coefficient ϕ and degree of reaction Λ:

$$\tan \beta_2 = \frac{1}{2\phi}\left(\frac{\psi}{2} - 2\Lambda\right) \tag{A3}$$

$$\tan \alpha_2 = \tan \beta_2 + \frac{1}{\phi} \tag{A4}$$

$$\tan \beta_3 = \frac{1}{2\phi}\left(\frac{\psi}{2} + 2\Lambda\right) \tag{A5}$$

$$\tan \alpha_3 = \tan \beta_3 - \frac{1}{\phi} \tag{A6}$$

From the calculated angles, the absolute and relative velocities can be found and thus the thermodynamic properties at the inlet and outlet of each stage can be determined.

$$C_2 = \frac{Ca}{\cos \alpha_2} \tag{A7}$$

$$C_3 = \frac{Ca}{\cos \alpha_3} \tag{A8}$$

$$V_2 = \frac{Ca}{\cos \beta_2} \tag{A9}$$

$$V_3 = \frac{Ca}{\cos \beta_3} \tag{A10}$$

Following from an initial estimate for nozzle loss coefficient the isentropic enthalpy (h'_2) can be calculated:

$$h'_2 = h_2 - \left[\frac{1}{2}\lambda_N C_2{}^2\right] \tag{A11}$$

where 1, 2 and 3 subscripts correspond to stator inlet, rotor inlet and rotor exit conditions respectively.

Using the the axial velocity, the area required to pass the specified mass-flow rate can be found. The passage area of the nozzle and rotor at different planes can be obtained as a function of the density ρ and axial velocity component.

$$A = \frac{\dot{m}}{\rho C_a} \tag{A12}$$

The blade height b and mean radius r_m can be calculated using the blade speed and rotational speed N:

$$r_m = \frac{U}{2\pi N} \tag{A13}$$

$$b = \frac{AN}{U} \tag{A14}$$

The stator and rotor passage losses can be obtained using the enthalpy loss coefficients. These coefficients are expressed as a function of the enthalpy difference between an isentropic and real expansion and the kinetic energy of the flow.

$$\lambda_N = \frac{h_2 - h_2'}{C^2/2} \tag{A15}$$

$$\lambda_R = \frac{h_3 - h_3''}{V^3/2} \tag{A16}$$

$$W = h_{03} - h_{01} \tag{A17}$$

From the estimated loss coefficients, the total-to-total isentropic efficiency can be determined.

$$\eta_{tt} = \left[1 + \left(\frac{\lambda_R \frac{V^3}{2} + \frac{C^2}{2}\lambda_N \frac{T_3}{T_2}}{h_{0_1} - h_{0_3}}\right)\right]^{-1} \tag{A18}$$

Following the above blade calculations, detailed dimensions for the rotor radius at the hub and tip can be obtained by applying the free vortex theory. Additionally, the number blades and blade profile can be obtained through estimating the optimum pitch to chord ratio and pitch to throat ratio as a function of the blade angles [28].

To check the design consistency with the permissible level of stress within the rotor blades the centrifugal stress applied on the blade and the gas bending stress can be calculated using the following equations assuming a tapered blade shape:

$$\sigma_{max} = \frac{4}{3}\pi N^2 \rho_b A \tag{A19}$$

$$\sigma_{bending} = \frac{\dot{m}Ca\left[\tan \alpha_2 + \tan \alpha_3\right]}{n_R} \times \frac{h}{2} \times \frac{1}{zc^3} \tag{A20}$$

where ρ_b is the density of blade material, n_R is the number of rotor blades, h is the mean blade height, A is the mean blade area and z is a constant obtained from a graph by Ainley [26].

References

1. Aichmayer, L.; Spelling, J.; Laumert, B. Thermoeconomic Analysis of a Solar Dish Micro Gas-turbine Combined-cycle Power Plant. *Energy Procedia* **2015**, *69*, 1089–1099. [CrossRef]
2. Yin, J.; Zheng, Q.; Peng, Z.; Zhang, X. Review of supercritical CO_2 power cycles integrated with CSP. *Int. J. Energy Res.* **2019**. [CrossRef]
3. Saeed, M.; Kim, K.H. Analysis of a recompression supercritical carbon dioxide power cycle with an integrated turbine design/optimization algorithm. *Energy* **2018**, *165*, 93–111. [CrossRef]
4. Moroz, L.; Frolov, B.; Burlaka, M.; Guriev, O. Turbomachinery Flow path Design and Performance Analysis for Supercritical CO_2. In Proceedings of the ASME Turbo Expo 2014, Düsseldorf, Germany, 16–20 June 2014. [CrossRef]
5. Holaind, N.; Bianchi, G.; De Miol, M.; Saravi, S.; Tassou, S.; Leroux, R.;Jouhara, H. Design of radial turbomachinery for supercritical CO_2 systems using theoretical and numerical CFD methodologies. *Energy Procedia* **2017**, *123*, 313–320. [CrossRef]
6. Qi, J.; Reddell, T.; Qin, K.; Hooman, K.;Jahn, I. Supercritical CO_2 Radial Turbine Design Performance as a Function of Turbine Size Parameters. *J. Turbomach.* **2017**, *139*. [CrossRef]
7. Zhou, A.; Song, J.; Li, X; Ren, X.; Gu, C. Aerodynamic design and numerical analysis of a radial inflow turbine for the supercritical carbon dioxide Brayton cycle. *Appl. Energy* **2017**, *205*, 187–209. [CrossRef]
8. Zhang, D.; Wang, Y.; Xie, Y. Investigation into Off-Design Performance of a sCO$_2$ Turbine Based on Concentrated Solar Power. *Energies* **2018**, *11*, 3014. [CrossRef]
9. Lv, G.; Yang, J.; Shao, W.; Wang, X. Aerodynamic design optimization of radial-inflow turbine in supercritical CO_2 cycles using a one-dimensional model. *Energy Convers. Manag.* **2018**, *165*, 827–839. [CrossRef]
10. Shi, D.; Zhang, L.; Xie, Y.; Zhang, D. Aerodynamic Design and Off-design Performance Analysis of a Multi-Stage S-CO2 Axial Turbine Based on Solar Power Generation Syste. *Appl. Sci.* **2019**, *9*, 714. [CrossRef]
11. White, M.T.; Sayma, A.I. A Preliminary Comparison of Different Turbine Architectures for A 100 Kw Supercritical CO_2 Rankine Cycle Turbine. In Proceedings of the 6th International Supercritical CO_2 Power Cycles Symposium, Pittsburgh, PA, USA, 27–29 March 2018.
12. Fiaschi, D.; Manfrida, G.; Maraschiello, F. Organic Rankine Cycles, ORC turbo expander design, Combined heat and power. *Appl. Energy* **2012**, *97*, 601–608. [CrossRef]
13. Casati, E.; Vitale, S.; Pini, M.; Persico, G.; Colonna, P. Centrifugal Turbines for Mini-Organic Rankine Cycle Power Systems. *J. Eng. Gas Turbines Power* **2014**, *136*. [CrossRef]
14. Rahbar, K.; Mahmoud, S.; Al-Dadah, R.; Nima, M. Parametric Analysis and Optimization of a Small-Scale Radial Turbine for Organic Rankine Cycle. *Energy* **2015**, *83*, 696–711. [CrossRef]
15. Da Lio, L.; Manente, G.; Lazzaretto, N. DA mean-line model to predict the design efficiency of radial inflow turbines in organic Rankine cycle (ORC) systems. *Appl. Therm. Eng.* **2018**, *132*, 245–255. [CrossRef]
16. Angelino, G. Carbon Dioxide Condensation Cycles For Power Production. *J. Eng. Power* **1968**, *90*, 287–295. [CrossRef]
17. Feher, E.G. The supercritical thermodynamic power cyclen. *Energy Convers.* **1968**, *8*, 85–90. [CrossRef]
18. Valencia-Chapi, R.; Coco-Enríquez, R.; Muñoz-Antón, J. Supercritical CO_2 Mixtures for Advanced Brayton Power Cycles in Line-Focusing Solar Power Plants. *Appl. Sci.* **2020**, *10*, 55. [CrossRef]
19. Binotti, M.; Invernizzi, C.; Iora, P.; Manzolini, G.; Dinitrogen tetroxide and carbon dioxide mixtures as working fluids in solar tower plants. *Sol. Energy* **2019**, *181*, 203–213. [CrossRef]
20. Manzolini, G.; Invernizzi, C.; Iora, P.; Binotti, M.; Bonalumi, D. CO2 mixtures as innovative working fluid in power cycles applied to solar plants. Techno-economic assessment. *Sol. Energy* **2019**, *181*, 530–544. [CrossRef]
21. SCARABEUS Project Home Page. Available online: http://www.scarabeusproject.eu/ (accessed on 26 November 2015).
22. Spadacini, C.; Pesatori, E.; Centemeri, L.; Lazzarin, N.; Macchi, R.; Sanvito, M. Optimized Cycle and Turbomachinery Configuration for an Intercooled, Recompression Sco2 Cycle. In Proceedings of the 6th International Supercritical CO_2 Power Cycles Symposium, Pittsburgh, PA, USA, 27–29 March 2018.
23. Lemmon, E.W.; Huber, M.L.; Mclinden, M.O. IST Standard Reference Database 23: Reference Fluid Thermodynamic and Transport Properties-REFPROP, Version 9.1. In Proceedings of the 6th National Institute of Standards and Technology, Standard Reference Data Program, Gaithersburg, MD, USA, 15–17 May 2013.

24. Simulis Thermodynamics. Available online: http://www.prosim.net/en/software-simulis-thermodynamics-mixture-properties-and-fluid-phase-equilibria-calculations-3.php (accessed on 13 November 2019).

25. Smith, S.F. A Simple Correlation of Turbine Efficiency. *J. R. Aeronaut. Soc.* **1965**, *69*, 467–470. [CrossRef]

26. Saravanamuttoo, H.; Rogers, G.; Cohen, H. Chapter 7—Axial and radial flow turbines. In *Gas Turbine Theory*, 5th ed.; Saravanamuttoo, H., Rogers, G., Cohen, H., Eds.; Pearson Education limited: Harlow, UK, 2001; pp. 305–366, ISBN 013015847X, 9780130158475.

27. Nanaware, A.; Pawar, S.; Ramachandran, M. Mechanical Characterization of Nickel Alloys on Turbine Blades. *REST J. Emerg. Trends Model. Manuf.* **2015**, *1*, 15–19.

28. Aungier, R.; Hal, C.A. Chapter 6—Preliminary Aerodynamic Design of Axial-Flow Turbine Stages. In *Turbine Aerodynamics: Axial-Flow and Radial-Flow Turbine Design and Analysis*; ASME Press: New York, NY, USA, 2005; pp. 133–165, ISBN 0791802418.

29. Dixon, S.L.; Hal, C.A. Chapter 4—Axial-Flow Turbines: Mean-Line Analysis and Design. In *Fluid Mechanics and Thermodynamics of Turbomachinery*, 5th ed.; Dixon, S.L.; Hal, C.A., Eds.; Butterworth-Heinemann: Boston, MA, USA, 2010; pp. 97–141, ISBN 978-1-85617-793-1.

30. Soderberg, C.R. *Unpublished Notes*; (quoted in reference [Dixon, 1989]); Gas Turbine Laboratory, Massachusetts Institute of Technolog: Cambridge, MA, USA, 1949.

31. Ainley, D.G.; Mathieson, G.C.R. *A Method of Performance Estimation for Axial-Flow Turbines*; H.M. Stationery Office: Richmond, UK, 1951.

32. Dunham, J.; Came, P.M. Improvements to the Ainley-Mathieson Method of Turbine Performance Prediction. *J. Eng. Power* **1970**, *92*, 252–256. [CrossRef]

33. Craig, H.R.M.; Cox, H.J.A. Performance Estimation of Axial Flow Turbines. *Proc. Inst. Mech. Eng.* **1970**, *185*, 407–424._PROC-1970-185-048-02. [CrossRef]

34. Kacker, S.C.; Okapuu, U. A Mean Line Prediction Method for Axial Flow Turbine Efficiency. *J. Eng. Power* **1982**, *104*, 111–119. [CrossRef]

35. Moustapha, S.H.; Kacker, S.C.; Tremblay, B. An Improved Incidence Losses Prediction Method for Turbine Airfoils. *J. Turbomach.* **1990**, *112*, 267–276. [CrossRef]

36. Horlock, J.H. Losses and efficiencies in axial-flow turbines. *Int. J. Mech. Sci.* **1960**, *2*, 48–75. [CrossRef]

37. Ahn, Y.; Bae, S.; Kim, M.; Cho, S.; Baik, S.; Lee, J.; Cha, J. Review of supercritical CO_2 power cycle technology and current status of research and development. *Nucl. Eng. Technol.* **2015**, *47*, 647–661. [CrossRef]

38. Wright, S.; Radel, R.; Vernon, M.; Rochau, G; Pickard, P. *Operation and Analysis of a Super-Critical CO_2 Brayton Cycle*; Sandia Report SAND2010-0171; Sandia National Laboratories Albuquerque, NM, USA, 2010. [CrossRef]

39. Clementoni, E.; Cox, T.; King, M. Off-Nominal Component Performance in a Supercritical Carbon Dioxide Brayton Cycle. *J. Eng. Gas Turbines Power* **2015**, *138*. [CrossRef]

40. Utamura, M.; Hasuike, H.; Ogawa, K.; Yamamoto, T.; Fukushima, T.; Watanabe, T.; Himeno, T. Demonstration of Supercritical CO_2 Closed Regenerative Brayton Cycle in a Bench Scale Experiment. In *Turbo Expo: Power for Land, Sea, and Air*; American Society of Mechanical Engineers: New York, NY, USA, 2012.

41. Schmitt, J.; Willis, R.; Amos, D.; Kapat, J.; Custer, C. Study of a Supercritical CO_2 Turbine with TIT of 1350 K for Brayton Cycle With 100 MW Class Output: Aerodynamic Analysis of Stage 1 Vane. In *Turbo Expo: Power for Land, Sea, and Air*; American Society of Mechanical Engineers: New York, NY, USA, 2014; Volume 3. [CrossRef]

42. Liu, K.; Lauwers, B.; Reynaerts, D. Process capabilities of Micro-EDM and its applications. *Int. J. Adv. Manuf. Technol.* **2007**, *47*, 11–19 doi:10.1007/s00170-009-2056-1. [CrossRef]

43. Cho, J.; Shin, H.; Ra, H.; and Lee, G.; Roh, C.; Lee, B.; Baik, Y. Development of the Supercritical Carbon Dioxide Power Cycle Experimental Loop in KIER. In Proceedings of the ASME Turbo Expo 2016, Seoul, Korea, 13–17 June 2016.
44. Moustapha, H.; Zelesky, M.F.; Balnes, N.C.; Japikse,D. Chapter 3—Preliminary and through flow design. In *Axial and Radial Turbines*; Concepts NREC: White River Junction, VT, USA, 2003; pp. 65–95, ISBN 978-0933283121.

 *applied sciences*

Article

Investigation of a Radial Turbine Design for a Utility-Scale Supercritical CO$_2$ Power Cycle

Tala El Samad [1], Joao Amaral Teixeira [2] and John Oakey [1,*]

[1]   Centre for Energy and Power, School of Water Energy and Environment, Cranfield University, Cranfield MK43 0AL, UK; T.ElSamad@cranfield.ac.uk
[2]   Centre for Propulsion Engineering, School of Aerospace Transport and Manufacturing, Cranfield University, Cranfield MK43 0AL, UK; J.A.Amaral.Teixeira@cranfield.ac.uk
*   Correspondence: J.E.Oakey@cranfield.ac.uk; Tel.: +44-7919-16-4688

Received: 26 May 2020; Accepted: 12 June 2020; Published: 17 June 2020

Abstract: This paper presents the design procedure and analysis of a radial turbine design for a mid-scale supercritical CO$_2$ power cycle. Firstly, thermodynamic analysis of a mid-range utility-scale cycle, similar to that proposed by NET Power, is established while lowering the turbine inlet temperature to 900 °C in order to remove cooling complexities within the radial turbine passages. The cycle conditions are then considered for the design of a 100 MW$_{th}$ power scale turbine by using lower and higher fidelity methods. A 510 mm diameter radial turbine, running at 21,409 rpm, capable of operating within a 5% range of the required cycle conditions, is designed and presented. Results from computational fluid dynamics simulations indicate the loss mechanisms responsible for the low-end value of the turbine total-to-total efficiency which is 69.87%. Those include shock losses at stator outlet, incidence losses at rotor inlet, and various mixing zones within the passage. Mechanical stress calculations show that the current blade design flow path of the rotor experiences tolerable stress values, however a more detailed re-visitation of disc design is necessitated to ensure an adequate safety margin for given materials. A discussion of the enabling technologies needed for the adoption of a mid-size radial turbine is given based on current advancements in seals, bearings, and materials for supercritical CO$_2$ cycles.

Keywords: supercritical carbon dioxide; radial turbine; utility-scale; turbomachinery design; NET Power

1. Introduction

In compliance with the warnings of the Intergovernmental Panel on Climate Change [1] to limit CO$_2$ discharges and keep the global temperature rise below 1.5 °C, techniques for mitigating power generation emissions are being widely investigated; examples range from using non-conventional working fluids and increasing the efficiency of power plants to implementing carbon capture and storage (CCS) [2]. Oxy-combustion appears to be the most favourable CCS route because of the simple separation of carbon dioxide from the steam present in the flue gases [3]. The basis of this method is that a fuel is combusted with pure oxygen to produce a stream of exhaust gases consisting mainly of H$_2$O and CO$_2$ which can be separated downstream through condensation [4].

With respect to unconventional working fluids, supercritical CO$_2$ (sCO$_2$) cycles are attracting growing interest due to the advantages associated with the fluid. The attractiveness of the use of sCO$_2$ fluid is based on its availability and inertness as well as the consistently cited advantages of such power cycles [5]:

- The nature of the high density sCO$_2$ fluid means that power conversion systems are very compact compared to conventional power cycles [6] ($\approx$30 times smaller than steam cycles and $\approx$6 times smaller than air cycles)

- High cycle efficiencies are also obtained, attributed to lower pumping power requirements and the non-ideal gas properties of sCO_2 which translate into lower fuel consumption and lower capital and operating costs. Targeted efficiencies of large-scale, closed supercritical CO_2 Brayton cycles approach 50%
- The suitability of the cycle for waste heat recovery from a variety of heat sources (e.g., nuclear, solar, fossil, geothermal, etc.)
- The application of these high-efficiency cycles reduces greenhouse gas effects because a CO_2 emission is used as a working fluid or as a recycle stream (i.e., not emitted)
- The suitability of sCO_2 cycle systems for use in arid climates because of their ability to use dry cooling (thus saving water) [7]
- Low purification requirements for fluid leakages because the cycle operates above the critical pressure of CO_2

The NET Power (Allam) cycle combines both originalities of oxy-combustion and supercritical CO_2 fluids in one system which is aimed at high power generation efficiencies coupled with high levels of CO_2 capture [8]. A review of oxy-combustion cycles has been reported by the International Energy Agency (IEA) identifying the main cycle configurations, requirements, benefits, complications and future developments by presenting cycle modelling and techno-economic study results [9]. The NET Power is one of the most attractive of the listed oxy-combustion cycles in terms of efficiency reaching 55% on lower heating value of fuel (LHV). Natural gas fuel or coal and oxygen from an air separation unit are combusted at high temperatures reaching 1200 °C at high pressures ($\approx$30 MPa) with the recirculated CO_2; due to the absence of N_2 in the combustion process, the flue gases are mostly composed of CO_2 and H_2O with no formation of NO_x (for a natural gas based cycle). The combustion products are then expanded in the CO_2 turbine with a pressure ratio between 6 and 12 before entering a multi-flow economizer heat exchanger where the turbine cooling flow and the recycled CO_2 and O_2 streams to the combustor are pre-heated using the residual flue gas heat [10]. The exhaust stream is also cooled for condensation and separation of water from the mixture. A fraction of the remaining (mostly pure) CO_2 stream is sent for carbon capture and storage while the rest is re-compressed and used as a temperature moderator in the combustor through mixing with the O_2 coming from the ASU [11]. Cooling of the combustor and turbine components to allow operation at the relatively high turbine inlet temperature of 1200 °C is more complex than in a conventional combined cycle gas turbine where cooling air can be bled directly from the compressor section. For the Allam cycle, the available cooling fluid is the CO_2 stream, unless a closed loop system is implemented using a different fluid, e.g., steam, although this would add cost as there is no steam available in the cycle. This constraint could limit the cycle from further efficiency improvements through increased turbine entry temperature due to the counterbalancing cooling penalty.

One of the most critical barriers that inhibits the full-scale development of the novel NET Power cycle is the design of the high-pressure, high temperature turbine which dictates cooling requirements, material considerations and number of stages to name a few. The turbine operates at an unorthodox combination of high temperatures, comparable to those of gas turbines, and high pressures, analogous to steam plants, in the presence of unconventional working fluids. The design of a proposed and undisclosed axial turbine, developed by Toshiba [12], requires intricate cooling passages within the small-sized blades, hence poses significant constraints on the design thereby limiting aerodynamic performance and manufacturability. A radial turbine, which generally has a simpler construction and fewer stages when compared to its axial contestant, is suggested as a superior candidate arrangement for such cycles of high fluid density. For these reasons, the NET Power theme is taken as the basis of a paradigm cycle configuration around which restrictions of a radial turbine design can be applied. A thermodynamic analysis of a mid-range cycle is established while lowering the turbine inlet temperature to remove cooling complications within the radial turbine passages. The cycle conditions are then considered for the design of a multi-MW scale turbine by using lower-order preliminary and higher fidelity methods.

The focus of this paper is the analysis of the results from a single satisfactory radial turbine design which is obtained based on the identified cycle requirements. It aims at highlighting the flow behaviour within the blade passages as opposed to solely trying to optimise a turbine design. Eventually, the methods are used to provide the necessary understanding of the loss mechanisms specific to sCO$_2$ radial turbines at such power scales, which in turn allow the identification of possible areas for re-design.

2. Modelling Methodology

The overall methodology followed throughout this piece of work is summarised in the flowchart of Figure 1. The main steps of the adopted modelling sequence are explained in the following Subsections. Initially, an uncooled cycle model is developed to obtain the turbine operating conditions required for the turbine design. The design procedure begins by performing initial calculations employing a meanline design code, and it is followed by generating a wide range of turbine geometries via preliminary design tools. The following stage involves the analysis of a set of the preliminary designs using 2D throughflow calculations where only a select few are taken further down the performance investigation using more advanced computational fluid dynamics (CFD) modelling. This is an iterative process until a suitable design that meets defined aerodynamic performance criteria is identified. Structural assessment is then performed for the chosen design to investigate the mechanical integrity of the rotor in operation.

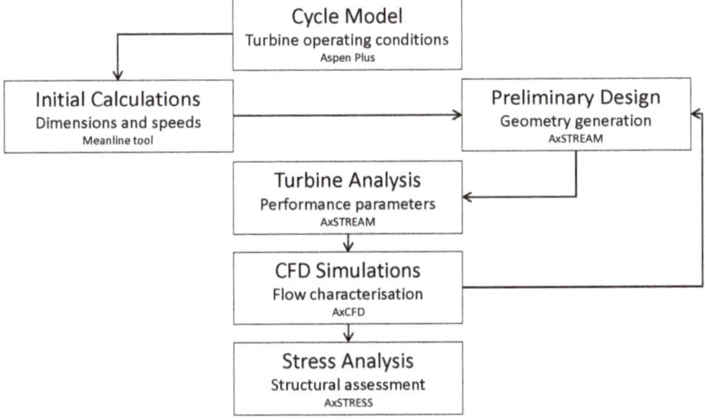

Figure 1. Overall methodology flowchart.

2.1. Cycle Modelling

The licensed software Aspen Plus V9 [13] is used for simulating the cycle, which is based on the initial model of the work of Scaccabarozzi et al. [10], to obtain the necessary data for turbine design which include working fluid ratios, power output, flow rates, as well as temperatures and pressures. Cycle modelling and analysis are performed before detailed geometrical information or structural aspects of components are defined.

The cycle considered in this work is a modification of the NET Power cycle introduced in Section 1. The changes include a reduced turbine inlet temperature and a lower pressure ratio; the uncooled turbine case is considered as a datum configuration for the following reasons:

- One of the benefits of a supercritical CO$_2$ cycle is small turbomachinery components which implies small blades that require intricate passages if cooling is employed. The design complexities associated with cooling are eliminated by the use of a smaller-scale radial turbine design
- Smaller power output scale so that a radial turbine can still be within acceptable efficiency ranges (power limit shown in Section 3.1.2)

- There is a lack of experience in radial turbine cooling for such high temperatures

The layout of the modified cycle under study is very similar to that in the work of Scaccabarozzi et al. [10]. Alterations to the original NET Power cycle model include the absence of cooling flows in the turbine (thus only one turbine component) and regenerator, plus the use of two flow compressors instead of four.

2.2. Meanline Design

The turbine operating conditions obtained from the cycle model are used in this intermediate step to give an insight on the possible performance range and geometric dimensions of the radial expander. A meanline design tool is developed using MATLAB by coding established turbomachinery equations.

The programme NIST REFPROP has been cited widely throughout the literature for defining the thermodynamic and transport properties in sCO_2 turbomachinery design tools [14–17]. It provides thermophysical properties of pure and compound fluids over a broad range of states for liquids, gases and supercritical phases based on validated data and calculation methods.

The initial design procedure is carried out using a meanline approach, which assumes one-dimensional passage conditions at mean radius of the turbomachine. The meanline aerodynamic design procedure of this work follows the approach introduced by Aungier [18] that can encompass the full operating range of a radial turbine. This method is used to determine the outline geometry of the turbine, given specific inputs, and to investigate the dependency of overall performance (size, efficiency, power) on key design parameters and assumptions such as specific speed and operating conditions [19]. Within the MATLAB tool, the estimate of the turbomachinery efficiency is obtained without the integration of loss models as these get incorporated in later stages of design. The main goal of this step is to obtain a range of turbine size parameters for use in the preliminary design tool of AxSTREAM in Section 2.3.

2.3. Preliminary Design

The proceeding design methodology tasks employ the integrated suite of turbomachinery design tools, AXSTREAM^{TM}, licensed by SoftInWay.

The preliminary design (PD) tool from AxSTREAM enables the fast computation to generate several possible turbine flow path designs for a set of given boundary conditions (pressure ratio, mass flow rate, range of rotational speed, flow coefficient) taking into account specified geometric constraints. The model follows a similar procedure to the meanline preliminary design process cited in Section 2.2, relying on the assignment of stage pressure drop and degree of reaction as independent variables of energy, continuity, state and process one-dimensional, steady, equilibrium and adiabatic equations. Firstly, velocity coefficients are refined to meet the criteria of the selected empirical design model, then losses are computed after obtaining a possible flow path, followed by the re-iteration of cascade angles for supersonic flows, and finally the definition of blade profiles is performed after obtaining flow angles.

The inverse task calculation of the preliminary design solves a set of equations, with the implementation of inlet and outlet boundary conditions, to search for the criteria of unknown flow angles as a function of generated design variables—flow coefficient and stage loading (both which relate to turbomachinery losses)—that are based on existing correlations. Details about the theoretical and mathematical background of the preliminary design tool can be found in [20]. Although there is a major three-dimensional aspect to the flow in a radial turbine, an approximate verification of 1D formulation is used in the PD step to provide an estimation of sizing data that depend heavily on the selected empirical loss methods [21].

The initial estimates obtained from the MATLAB meanline code (without loss models) are used as input data to AxSTREAM. The preliminary design tool generates a large pool of design solutions based on the given boundary data and geometric constraints. The particular license of the software employed does not provide the capability of having a fluid mixture (such as that present in the cycle) so the working fluid is taken as pure supercritical CO_2 assigned by the AxS Carbon Dioxide property method which is a modified Redlich-Kwong equation of state that covers a wide range of pressures and enthalpies [22]. Default empirical methods and loss models for radial turbines embedded in AxSTREAM are used in the design tasks.

All the generated turbine solutions can be explored in depth in the counterpart of the PD tool of AxSTREAM, the design space explorer, where designs are filtered through by applying constraints and limitations to allow trade-offs between power, efficiency, geometric design and velocities. Table 1 shows the filters applied to the design space to narrow the results selection process. The turbine solutions that do not meet the pre-defined constraints are automatically discarded and the selected designs can be examined in terms of flow path geometry. This option also allows for a quick study of the influence of crucial parameters on performance and size at the early stages of preliminary design.

Table 1. Design space explorer constraints.

Parameter	Unit	Constraint
Power	MW	$\geq \dot{W}$ from cycle Section 3.1
Efficiency	%	≥ 80
Flow coefficient	-	<1
Rotor inlet blade speed	m/s	≈ 450
Rotor outlet absolute flow angle	°	≈ 0

2.4. Turbine Throughflow Analysis

The selected turbines from the preliminary design step are then analysed using the throughflow tool of AxSTREAM. Meanline analysis can be performed by using only the mean curvature section of the blade. Whereas a more refined streamline benchmarking calculation is done using the direct solver that considers the cascade geometry at hub, mid-span and tip for both stator and rotor. Kinematic and thermodynamic parameters are obtained along the blade height at three sections (hub, mid-span and tip). The aim is to get more precise performance parameters of the turbine before proceeding to blade profiling and design if needed.

Once a number of results are obtained for a select number of turbines, the user can start the post-design process of reviewing and editing the turbine flow path if needed; turbines that have not shown good performance or have not achieved solution convergence are discarded at this step instead of trying to optimise and edit the design. The outcome of the streamline calculations include loss and velocity charts, updated velocity triangles, meridional flow path and the enthalpy- entropy diagram, all which can be saved to the main project database to override performance results from the PD step. The project database will hold all relevant turbine information for each section (hub, mean, tip) as well as details on geometric and loss values for the design. Another check on the required constraints is performed in case any design fails to meet the criteria so as not carry it further to higher fidelity simulations.

2.5. Computational Fluid Dynamics

AxSTREAM provides tools that generate geometries based on 1D meanline design and allows analysis using higher fidelity models ranging from streamline analysis, axi-symmetric 2D and 3D computational fluid dynamics (CFD). AxCFD is the tool used for pre-processing, mesh generation, calculation and results post-processing. This step allows for the evaluation of the designed flow path, the comparison of results between the different calculation models (meanline, streamline, 3D CFD), and the understanding of flow mechanisms within the turbine.

The turbine domain employing a mixing plane halfway between stator trailing edge and rotor leading edge is simulated using AxCFD. Boundary conditions, of $P_{0_{in}} = 297$ bar, $T_{0_{in}} = 1173$ K and $P_{out} = 54.0$ bar, and working parameters are taken directly from the AxSTREAM project case saved after the 2D analysis with values taken as spanwise averages of quantities at hub, mean and tip sections. Rotational speed and mass flow rate are also automatically inherited from the project database but are altered for off-design simulations. Like other CFD tools, AxCFD solves decomposed Reynolds-Stress-Averaged Navier Stokes (RANS) equations with time-averaged density, pressure and energy conservation, and mass-averaged velocity equations in discretised fluid domains.

Passage mesh creation is required before any aerodynamic assessment of the turbine. The way of defining the size and number of the mesh elements in AxCFD is through mesh quality, a value ranging between 1 (coarsest) and 10 (finest). A mesh sensitivity study is performed, with results discussed in Section 3.3.1, to allow for a compromise between accuracy (mesh refinement) and computational time. Structured HO-type mesh elements with the same quality in hub-to-shroud and blade-to-blade directions are used in all cases with additional parameters for mesh adjustment; values are taken based on recommendations by SoftInWay.

The meshing zones defined by points on leading edge (LE) and trailing edge (TE) are modified manually over the blade profile contour for hub and tip sections. A mesh quality check is performed; good quality elements are described as those with no Jacobian determinant less than 0.3 and no (or minimal) angle parameter less than 12°.

The k-ω SST turbulence model is used for its suitability in viscous turbomachinery simulations based on previous experience [23–25], with a medium turbulence intensity of 5%. The first order upwind scheme is employed due to convergence difficulties when using second-order. Slightly fluctuating trends and considerably high value criteria for some residuals in the range of 10^{-5} are accepted for simulations of similar applications shown in [26,27], but convergence is also checked against inlet/outlet mass imbalance of within 0.2%. AxCFD also allows calculations over 2D spanwise fluid domain sections via axi-symmetric modelling which is used of off-design performance map generation.

3. Results

3.1. Cycle Modelling

Before proceeding to process modelling of the fully modified cycle and using it as a basis for turbine design, it is reasonable to try and understand if there are any cycle benefits which can be credited to the removal of turbine cooling beyond that of reducing design complexities. From a practical point, a comparison between thermodynamically-identical cycles is conducted where cooling is the only limiting distinction.

3.1.1. Cooled Cycle Comparison

In an attempt to compare the performance of an uncooled cycle (the target cycle of this work) to a cooled one, like-for-like working conditions are applied. Conditions of the optimised cycle model from the work of Scaccabarozzi et al. are adopted [10] which are compared against the data from IEA GHG report [9]. Optimisation work, to achieve maximum cycle efficiency target, was carried out by the authors of [10] after performing a sensitivity analysis on the most relevant cycle-affecting parameters. The thermal energy of the feedstock is kept the same as in the cited works by setting the natural gas inlet flow rate to 16.52 kg/s. For the sake of comparison, the uncooled turbine is assumed to be capable of operating at a full-size ($\approx$700–800 MW) power scale.

The comparison between a number of cycle performance results is viewed in Table 2. Generally there is good agreement in the results with some discrepancies that can be attributed to the differences in minor quantitative assumptions, and because of using two compression stages instead of four [10] which lowers the power required to drive the compressors. The absence of turbine cooling gives a higher turbine outlet temperature which implies a lower temperature drop across the turbine contributing to a lower turbine specific work. However, the higher combustion flow rate, required to reach the desired temperature of the cycle, balances the drop in turbine specific work. Overall, the cycle with the uncooled turbine gives a superior cycle efficiency compared to the cooled case, a 1.54% efficiency point gain. The heat transfer associated to preheating the turbine cooling flow within the regenerator (which also contributes to expansion losses) is removed and thus the detrimental efficiency penalty affiliated with this process is eliminated causing a beneficial impact on the cycle efficiency.

In fact it is suggested that while turbine cooling allows an increase in turbine inlet temperature (TIT) which enhances cycle performance, it also increases energy losses in turbines and thus leads to a lower aerodynamic efficiency [28] that in turn would have a great negative impact on the net cycle efficiency [29]. Scaccabarozzi et al. [10] noted that the maximum net cycle efficiency is achieved at a relatively low turbine inlet temperature (lower than the maximum simulated TIT value) because of the substantial cooling mass flows required if a high combustor outlet temperature is implemented.

Table 2. Comparison of optimised cycle between [10] and uncooled turbine model.

	Unit	Results of This Work	Results of [10] Optimised Cycle
Thermal energy of feedstock (LHV)	MW_{th}	768.21	768.31
Turbine power output	MW_e	610.55	609.74
Recycle flow compressors	MW_e	88.00	97.81
Natural gas compressor	MW_e	3.97	-
Air separation unit	MW_e	85.51	-
Storage compressor	MW_e	0.25	-
Net electric power output	MW_e	432.82	421.06
Turbine outlet temperature	°C	830.53	783.81
Turbine inlet flow rate	kg/s	1567.7	1513.7
Net electric efficiency (LHV)	%	56.34	54.80

3.1.2. Modified Cycle

As discussed in Section 1, supercritical CO_2 cycles, regardless of whether they employ cooling or not, have the beneficial characteristics of small turbomachinery components due to the high fluid density. Suggestions from literature limit the applicability of radial turbines to cycles of up to 30 MW_e [30]. The criteria for that power level was proposed on the basis of closed sCO$_2$ cycles with lower operating pressures and temperature compared to conditions witnessed in cycles similar to the Allam cycle.

The numerical value of a turbine specific speed is used as an index of power output and passage size [31]; the dimensionless parameter is used by designers to determine turbine type and efficiency. Dixon shows that for a specific speed $\left(n_s = \dfrac{\omega\sqrt{\dot{Q}_5}}{(\Delta h_{id})^{0.75}} \right)$ in the narrow range of 0.3–1.0, where this span corresponds to small turbine passage areas, a 90° inward-flow radial (IFR) turbine can reach high efficiencies in contrast with the conventional axial turbines which have a much broader spectrum of peak performance [31].

A larger-scale turbine—higher volumetric ($\dot{Q}$) or mass flow rate ($\dot{m}$)—corresponds to a higher specific speed value for fixed ideal heat drops (Δh_{id}) and rotational speeds (ω). Thus, taking a within-the-range recommended value of n_s = 0.7 [18,31] and knowing the ideal enthalpy drop from the cycle model (Δh_{id} = 284,370 J/kg), plus having performed some calculations on possible rotational speeds being in the range of 25,000 rpm, through initial turbine sizing (i.e., ω = 2618 rad/s) the corresponding maximum allowable mass flow rate is:

$$\dot{m}_{max} = \dot{Q}_{max} \cdot \rho_{out} = \left(\frac{0.7 \Delta h_{id}^{0.75}}{\omega} \right)^2 \cdot \rho_{out} \approx 354 \text{ kg/s} \tag{1}$$

where ρ_{out} is the density of the flow at turbine outlet and is around 33.2 kg/m^3 from the cycle model.

The limit obtained for the largest size of a radial turbine operating under the conditions of a dense working environment, at TIT = 900 °C, turbine inlet pressure (TIP) = 297 bar and pressure ratio

(Π) = 4.95, is used to assign the scale of the cycle and assess the applicability of such a size. The cycle model is altered with the selection of $\dot{m}_{max}$ and an assumed medium-value turbine isentropic efficiency of 85%, and the final modified cycle performance parameters are listed in Table 3. Although the net electric power output of 67.33 MW_e is not within the scale range of the suggested 700 MW NET Power plans, the net work of this cycle is in the scale of gas turbines used in small industrial and commercial applications by major OEMs. The market for turbines in the range of less than or equal to 70 MW_e is believed to grow considerably in the near future [32].

To summarise, Table 4 offers the turbine boundary conditions which are used to design, size and assess a radial expander that could be fit for use. The obtained turbine fluid components is comparable to the NET Power working environment reported in literature [9] with a slightly higher CO_2 mole fraction. This is because in the original NET Power cycle the combustion products are mixed with the turbine cooling stream which has a high CO_2 component. Although the major fluid component entering the turbine is CO_2 with 94.06% mole fraction, there are other fluid contaminants that could affect turbine design and performance. However, the assumption of a pure CO_2 fluid is employed throughout the design process.

Table 3. Modified cycle performance parameters.

Data	Unit	Value
HTHE heat duty	MW	242.92
LTHE heat duty	MW	43.42
Thermal energy of feedstock (LHV)	MW_{th}	137.18
Turbine power output	MW_e	99.31
Vapor phase compressors	MW_e	3.96
Dense phase compressors	MW_e	8.86
Oxidant compressor	MW_e	3.09
Total flow compressors	MW_e	15.91
Natural gas compressor	MW_e	0.74
Air separation unit	MW_e	15.29
Storage compressor	MW_e	0.045
Net electric power output	MW_e	67.33
Net electric efficiency (LHV)	%	49.08

Table 4. Radial turbine boundary conditions from cycle model.

Data	Unit	Value
Inlet temperature (T)	°C	900
Inlet pressure (P)	bar	297
Mass flow rate	kg/s	354.2
Outlet pressure	bar	60
Thermal power output ($\dot{W}$)	MW_{th}	100.7
Fluid Component Mole Fractions		
CO_2	%	94.06
H_2O	%	4.13
N_2	%	1.14
Ar	%	0.55
O_2	%	0.12

3.2. Preliminary Design

The design phase that builds on the results of the cycle analysis and the meanline code consists of the definition of three-dimensional geometries which are then further investigated using higher fidelity tools. The tool employed in this phase is the preliminary design solution generator, which is part of the AxSTREAM design suite, is initiated with defined settings (working fluid, loss models, number of stages), boundary conditions and geometry constraints. The boundary conditions required to obtain the preliminary design solutions are based on design specifications from the cycle requirements that are fixed, whereas all other parameters obtained from an initial meanline calculation are specified as a range of values. The particular range employed is defined based on the survey of literature together with outcomes of the sensitivity analysis from the initial calculations step. The inlet total temperature and outlet total pressure are used as boundary conditions and the criterion against which output designs are selected is based on the total-to-static efficiency. A vaned nozzle is used and based on results not shown in this paper, cases without volutes are selected which means that an annular collector would be required to gather the inlet flow to the turbine. The rotor inlet metal angle is set to 90° (i.e., straight radial blades at inlet) for manufacturability and structural reasons. Full input data generated from the meanline design code is found in Table 5.

Table 5. Input data for preliminary turbine design.

Variable	Unit	Value
Boundary Conditions		
Inlet total pressure	bar	297
Inlet total temperature	K	1173
Outlet total pressure	bar	60
Mass flow rate	kg/s	354.2
Inlet rotor metal angle	°	90
Shaft rotational speed	rpm	17,000–24,000
Geometric Parameters and Constraints		
Stator outlet mean diameter	m	0.4–0.6
Rotor blade height ratio	-	2–3
Rotor diameter ratio	-	1.5–3
Rotor inlet height	m	0.01–0.04
Hub reaction	-	0.3–0.5
Stator inlet diameter	m	0.4–0.8
Inlet flow angle	°	1–179
Rotor outlet hub diameter	m	0.05–0.3

The preliminary design space allows the visualisation of the estimated set of turbine solutions for the given input data. An example of a generated design space with 130 possible solutions is shown in Figure 2 with each point corresponding to a particular turbine design. The total-to-static efficiency is represented by the colour of each point for both plots. Higher efficiency values coincide with the lower range of flow coefficients and higher velocity ratios (faster rotational speeds). In some cases, a qualification between efficiency and velocity ratio is needed to limit the induced rotor tip speed from high rotational speeds because of permissible material limits. Greater power is produced from the designs with higher efficiencies as expected.

From the preliminary design space, designs are filtered through firstly by imposing general constraints within the design space explorer that automatically discards points, followed by manually selecting the turbines and assessing geometries and velocity triangles. The design space explorer is used to set constraints to narrow down the selection process, and to compare designs that meet the turbine requirements. Filter limits are specified to automatically and rapidly remove unsatisfactory points; the imposed criteria is to discard any turbines that have $\dot{W} < 100$ MW and efficiencies $< 80\%$. 78 out of the 130 solutions do not meet the imposed criteria and are removed from the selection process. The remaining solutions are individually selected for further assessment of the general geometric features and velocity triangles in order to ensure that the restrictions on tip speeds and flow angles are applied.

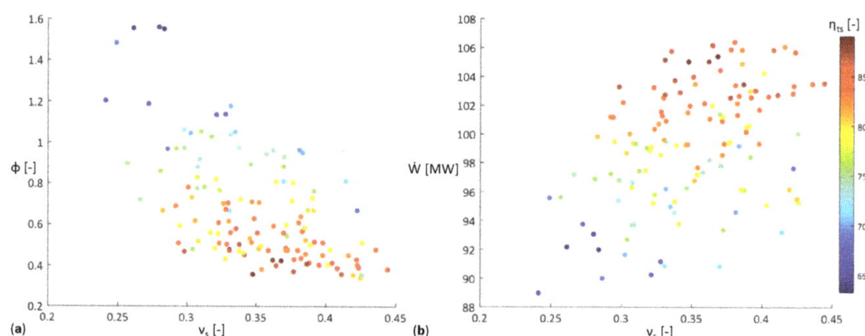

Figure 2. Preliminary design space: (**a**) Flow coefficient vs velocity ratio, (**b**) Power vs velocity ratio.

From the few selected designs, some are discarded because they either exhibit high circumferential rotor blade tip speed, or did not meet performance metrics, or the exit absolute flow angle was far from zero, or they had narrow passages which would incur high blockage. One particular design that meets the defined criteria, and which is characterised by a smooth rotor turning from the radial to the axial direction is presented in Figure 3, with the meridional dimensions given in Figure 4. Additional geometric parameters of the radial turbine are found in Table 6; the blade metal angle (θ) is defined as the angle between the tangential reference plane and the blade camberline extension at leading edge for inlet, and trailing edge for outlet. In the case of no volute, the flow at vane inlet is assumed almost purely radial with straight radial blades at rotor inlet as well ($\theta_{in} \approx 90°$). The shaft rotational speed is 21,409 rpm and the rotor tip clearance is 0.782 mm.

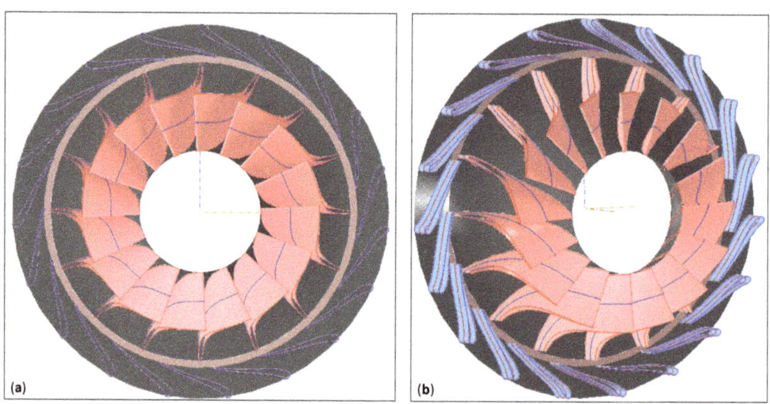

Figure 3. 3D model of selected turbine: (**a**) Front view, (**b**) Perspective view.

Table 6. Geometric parameters of turbine stator ($_S$) and rotor ($_R$) blades.

Stator			Rotor		
Variable	**Unit**	**Value**	**Data**	**Unit**	**Value**
Blade count (N_S)	-	19	N_R	-	16
Chord (c_S)	mm	99.90	c_R	mm	176.0
Pitch (q_S)	mm	70.24	q_{R_m}	mm	47.81
Inlet thickness (t_2)	mm	9.150	t_4	mm	3.520
Outlet thickness (t_3)	mm	1.830	t_5	mm	3.520
Throat (o_S)	mm	16.82	o_R	mm	28.35
Inlet angle (θ_2)	°	112.42	θ_4	°	90
Outlet angle (θ_3)	°	13.86	θ_{5_m}	°	36.37

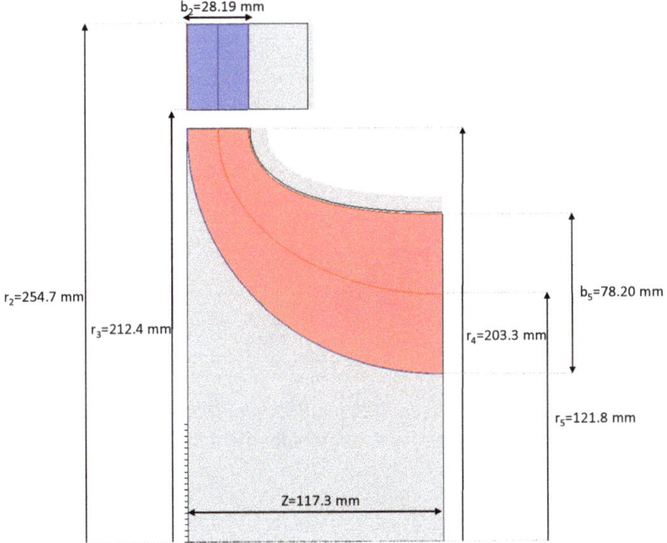

Figure 4. Meridional dimensions of selected turbine.

The performance parameters obtained upon completion of 2D throughflow analysis for the turbine case are listed in Table 7. These results are obtained assuming inviscid flow.

Table 7. Anticipated performance results.

Variable	Unit	Value
$\dot{m}$	kg/s	354.6
$\dot{W}$	MW	101.6
Total-to-total efficiency (η_{tt})	%	88.25
Total-to-static efficiency (η_{ts})	%	83.40
ν_s	–	0.344
Stage loading (ψ)	–	1.42
ϕ	–	0.433
Degree of reaction (R)	–	0.282
Π_{ts}	–	5.52

3.3. Computational Fluid Dynamics Analyses

Three-dimensional simulations are carried out for a single blade passage of the turbine to form a basis of obtaining performance results to be contrasted against lower-fidelity analysis techniques, and acquiring an insight into the turbine flow which can be used for comparison purposes with other designs (not presented in this paper), given that there is no available validation data for designs similar to the radial turbine under examination.

3.3.1. Mesh Sensitivity

A mesh sensitivity study is performed to examine the relationship between computational time, which is related to total nodal count, and convergence of results. Two sets of boundary conditions definition are imposed to check discrepancies between mesh densities with results listed in Table 8. The largest observed changes refer to the efficiency values which can be imputed to improved wall boundary layer resolutions as number of elements increase.

Table 8. Performance results for 3D CFD mesh sensitivity study using two different boundary condition definitions.

Boundary Condition: $\dot{m}_{in}$ & $\mathbf{P}_{out}$				
Number of Elements [-]	$\mathbf{P}_{0_{in}}$ **[bar]**	$\dot{W}$ **[MW]**	η_{tt} **[%]**	η_{ts} **[%]**
51,110	224	83.97	72.66	67.50
209,308	281	93.16	73.48	67.98
971,613	282	102.2	69.77	63.17

Boundary Condition: $\mathbf{P}_{0_{in}}$ & $\mathbf{P}_{out}$				
Number of Elements [-]	$\dot{m}$ **[kg/s]**	$\dot{W}$ **[MW]**	η_{tt} **[%]**	η_{ts} **[%]**
51,110	377.3	81.48	76.28	70.42
209,308	375.2	101.3	73.01	67.21
971,613	372.5	108.5	69.87	63.05

The finest mesh of 971,613 elements is used for flow visualisation. The y^+ on the rotor surfaces is ≤ 2; it is harder to obtain low values of the non-dimensional distance on the stator and rotor shroud because of the high Reynolds number (Re $\approx 1.7 \times 10^7$, given that the fluid has a high density). The range of y^+ values on all surfaces is between 0.14–200 (200 being the highest acceptable limit of y^+ as referred to SoftInWay manual recommendations). Cerdoun et al. [33] also noted high values of y^+ in specific regions of the CFD mesh (0.3–92.2). Lower y^+ can be obtained by further increasing mesh element count but owing to the licensing restrictions of AxCFD and the limited computational resources available to run the simulations, the mentioned values are accepted. H-O grid topologies are used in both the stator and rotor domains for smooth alignment with the flow direction, and refinement made in regions near blade walls and leading/trailing edges to account for high normal gradients [34].

Inlet/outlet mass imbalance, monitoring of total-to-total efficiency stabilisation, and root mean square residuals of mass, density, pressure and turbulence equations (k and ω) are used as criteria to assess solution convergence. For lower mesh densities, the residual values are set at $\leq 10^{-6}$ but for finer meshes, residuals require more iterations to reach steady higher values. This is observed in similar works by Wei [26] where increased residual values are accepted for off-design conditions and for finer mesh simulations.

3.3.2. Design Point Turbine Analysis

Performance data obtained from the numerical 3D turbulent simulations is listed in Table 9 which can be compared to the 2D analysis results found in Table 7. Global variables of mass flow, power, and non-dimensional parameters all show good agreement; the small difference in thermodynamic properties between the analyses outcomes (enthalpies and temperatures) lead to a greater difference in power output. The total pressure at turbine outlet is 62.9 bar with a total temperature of 975 K which are both slightly higher than the values initially anticipated. The most pronounced discrepancy can be observed in the values of efficiencies that can reach 20%. This change is not fully uncommon with comparisons of 15% efficiency difference between meanline design and CFD results noted by Sauret et al. [35] in their design of a R134A radial turbine, and almost 24% efficiency difference reported by Meijboom [23] between 1D design and 3D CFD modelling for a supercritical CO_2 radial turbine. The source of this variance can be attributed to a number of factors, the turbine 3D blade geometries not being fully optimised, and the effects of three-dimensional viscous losses and tip clearance not accounted for in earlier analysis stages. The current design corresponds to an intermediate development stage and requires further adjustment based on the following observations made hereafter.

Table 9. 3D CFD performance results.

Variable	Unit	Value
$\dot{m}$	kg/s	372.5
$\dot{W}$	MW	108.5
η_{tt}	%	69.87
η_{ts}	%	63.05
ψ	-	1.34
ϕ	-	0.511
R	-	0.355
Π_{ts}	-	5.51

Figure 5 shows relative Mach number (M_{rel}) distribution (absolute for stator) at three locations across the span. At the throat of the stator, the flow reaches its supersonic conditions with the onset of the trailing edge shock; high losses are associated with shockwaves which are mitigated by changing airfoil shape to lower Mach number values. Low Mach number values near the rotor pressure side (PS) leading edge is caused by incorrect incidence and the propagating suction side acceleration interacting with the presence of the rotor tip clearance; the area of incidence loss near the pressure side lessens from 10% to 90% span with the mean relative flow angle at rotor inlet $\beta_4 = 28.22°$ which is lower than the calculated flow angle from throughflow analysis ($\beta_4 = 45.50°$). The low M_{rel} values on the suction side indicate a region of flow separation that begins at SS leading edge and grows along the flow as end-span is approached (also reported in Dong et al.'s work [24]).

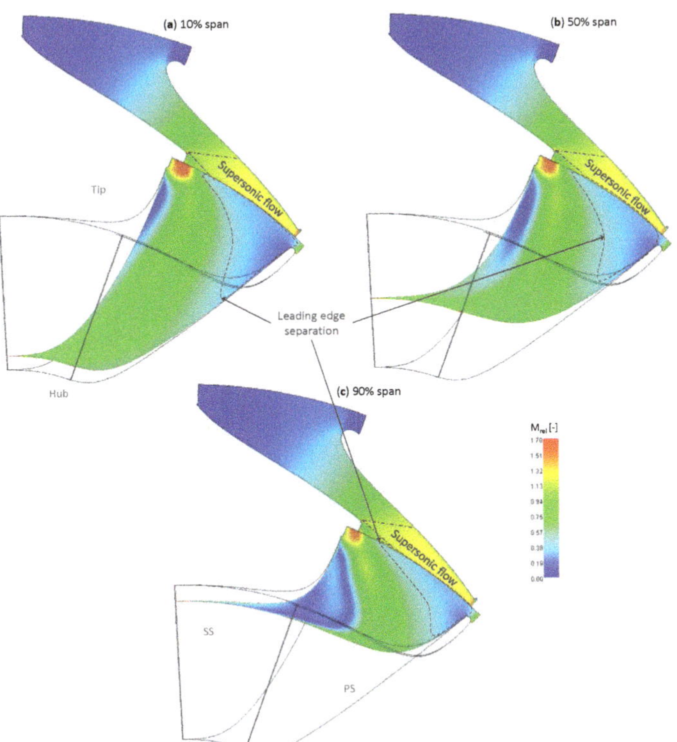

Figure 5. Relative Mach number distribution at three spanwise turbine passage locations: (**a**) 10% span, (**b**) 50% span, (**c**) 90% span.

Figures 6 and 7 can be viewed in conjunction to further investigate loss zones in the radial turbine rotor passage. This blade-to-blade view shows more detail regarding the flow characteristics on the suction and pressure surface of the rotor blade and omits details of stator and rotor leading edge flows. Boundary layer separation is observed towards the end of the span just after the rotor blade angle experiences a major deflection on the suction side; the separation zone in Figure 6 corresponds to the high entropy generation area in Figure 7. Smaller regions of entropy rise are noticed near the leading edge because of the reattachment of the leading edge separation (because of incorrect incidence) on the pressure side with the higher velocity flow on the suction side. Rotor wakes, produced by the velocity differential between the viscous layers at trailing edge from the suction and pressure sides, also correspond to higher irreversibility especially at mid-span. The severity of the loss locations and their expanse help explain the low efficiency obtained for the turbine design. The entropy rise across the passage is more than double the initially predicted value (88.92 J/kg K vs 41.07 J/kg K). The flow deceleration in some parts of the passage indicate separation or flow reversal. Similar flow behaviour patterns have been remarked in works on radial turbines and turbochargers performed by other researchers. Wei [26] reported separation on the blade suction side for a supercritical CO_2 radial turbine and Rahbar et al. [19] outlined leading edge separation on the pressure side depending on rotor blade design and turning.

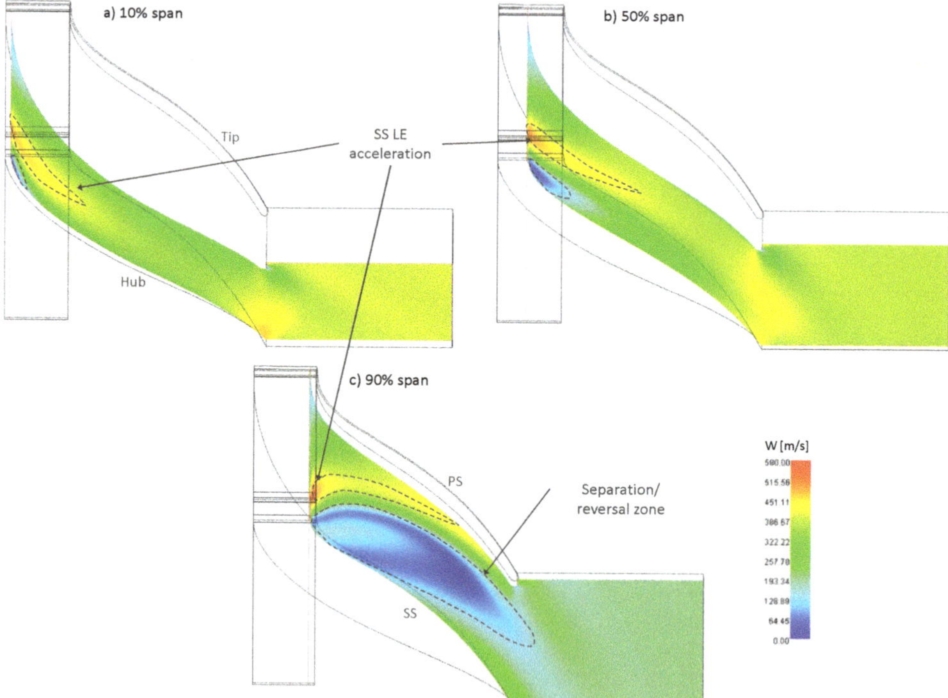

Figure 6. Relative velocity contour at three spansiwse rotor passage locations: (**a**) 10% span, (**b**) 50% span, (**c**) 90% span.

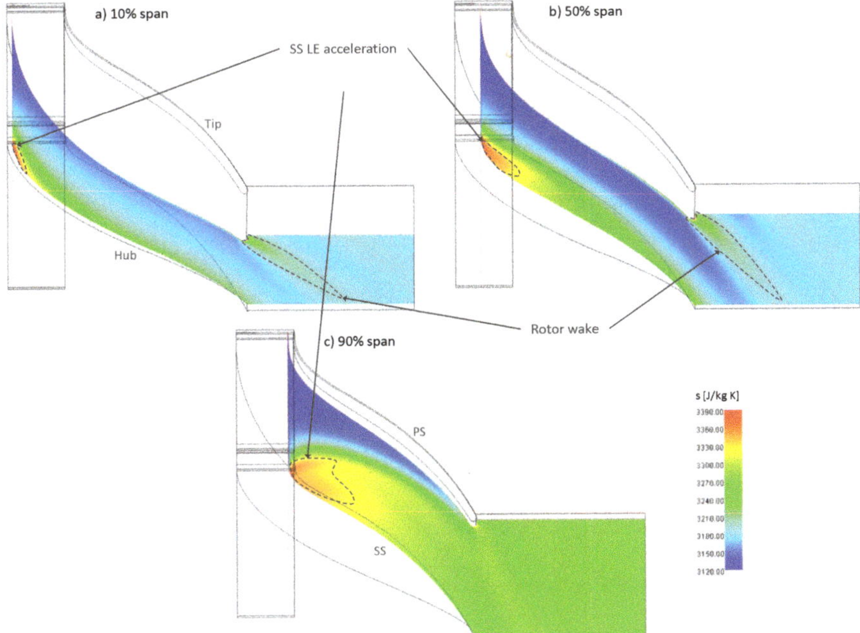

Figure 7. Entropy contours at three spanwise rotor passage locations: (**a**) 10% span, (**b**) 50% span, (**c**) 90% span.

Total pressure losses occur in the rotor passage as indicated in Figure 8. At sections after rotor entry, the most pronounced total pressure loss (blue colour) happens on the suction side from hub to tip because the pressure side still maintains the high pressure characteristic. As the flow moves downstream, the total pressure loss vortex loses its magnitude compared to the bulk flow (green colour with no gradient), shifts towards the shroud region, and extends to reach the pressure side near rotor exit because of the formation of tip leakage.

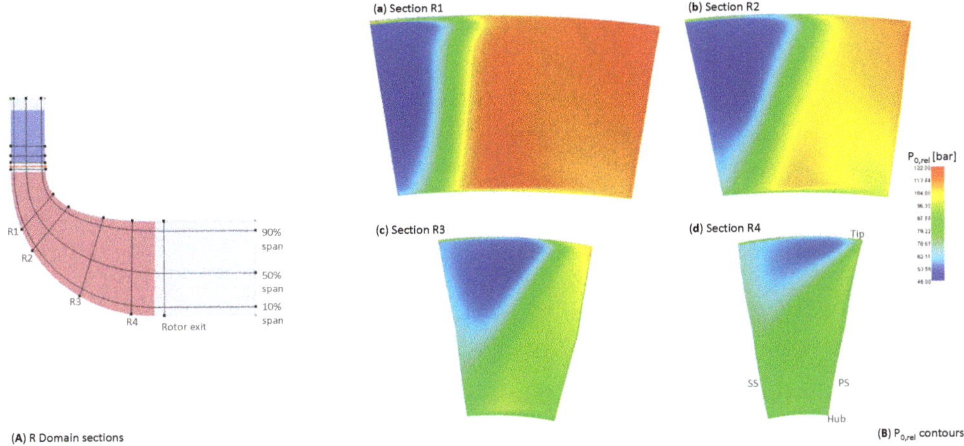

Figure 8. (**A**) Domain stream definition, (**B**) Relative rotor pressure distribution in rotor stream-sections, flowing downstream: (**a**) R1, (**b**) R2, (**c**) R3, (**c**) R4.

An additional plane at rotor exit with contours of total pressure in the rotating frame of reference is plotted in Figure 9. The warm-coloured patches (regions tending to red) are areas with the highest total pressure, whilst cooler-coloured tones (tending to blue) are zones where total pressure is lost. The large tip clearance leakage area, visible at the rotor exit stream-section, adds to the amount of entropy/loss generation which leads to low efficiency values.

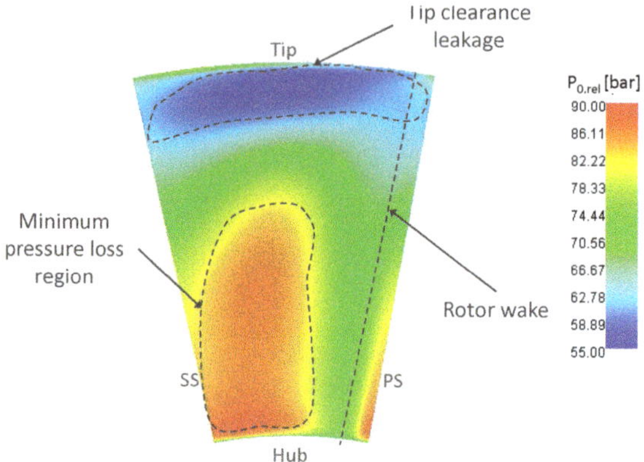

Figure 9. Relative total pressure contours at rotor exit stream section.

3.3.3. Off-Design Turbine Analysis

The turbine performance maps are generated using 2D CFD simulations rather than the full 3D ones, on account of the computational resources available as well as the long time required for the 3D simulations. The conditions extend over the range of total-to-static pressure ratio between 2 and 7 at three rotational speeds of 19,268 rpm, 21,409 rpm and 23,550 rpm. The variation of mass flow rate and stator outlet Mach number with the off-design conditions is shown in Figure 10. In general, the mass flow is expected to increase with higher pressure ratios but have lower values at higher rotational speeds, until choking occurs whereby the mass flow rate reaches its maximum asymptote. The radial turbine becomes choked at around $\Pi_{ts} = 3$, thus for the operating range of the design, the turbine will almost always experience supersonic flows.

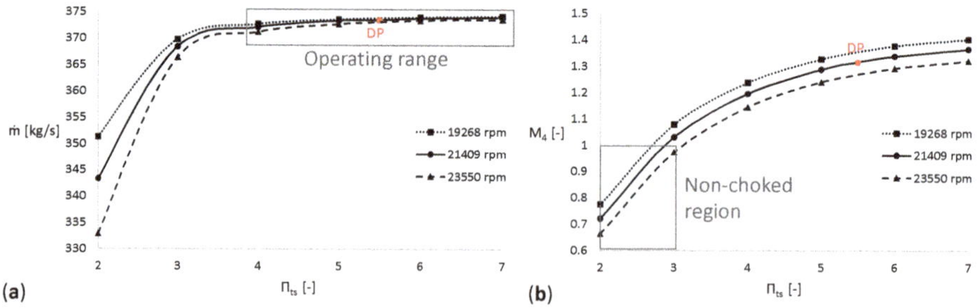

Figure 10. Variation of parameters with total-to-static pressure ratio at three rotational speeds: (**a**) Mass flow, (**b**) Stator outlet Mach number.

Both total-to-total and total-to-static efficiency maps are plotted in Figure 11. The difference between the efficiency values is solely attributed to the loss in the available kinetic energy at rotor outlet, i.e., related to the exit swirl (α_5), because the axial velocity is not a function of rotational speed (but is related to pressure ratio). The trend for both efficiencies are broadly similar except for the value difference due to the mentioned loss. The highest efficiency value occurs with a pressure ratio of about

3 at 23,550 rpm and the lowest with $\Pi_{ts} = 7$ at 29,268 rpm. The maximum Mach number at stator outlet is lower for higher ω (as deduced from Figure 10) meaning that the entropy rise in the stator exit channel is also lower, hence the higher efficiencies.

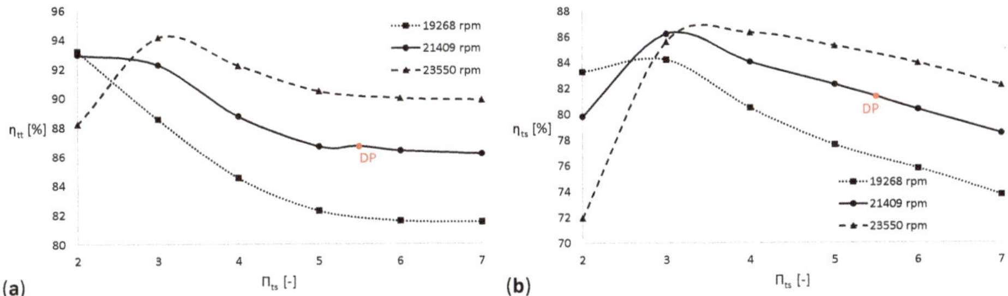

Figure 11. Variation of efficiencies with total-to-static pressure ratio at three rotational speeds: (**a**) Total-to-total efficiency, (**b**) Total-to-static efficiency.

Table 10 shows the change in relative flow angle at rotor inlet and absolute flow angle at rotor outlet at the off-design conditions. The flow angle at rotor inlet (β_4) plays a role in determining incidence losses. The recommended values of rotor incidence angles for a radial turbine are in the range of $-40°$–$-20°$ [36]; incidence angles around the suggested values are only obtained in two cases ($-37.01°$ & $-44.19°$), all which correspond to pressure ratios below choking conditions and those cases correspond to relatively high efficiencies. Ideally, the closer the absolute flow angle at rotor outlet is to zero, the lower the kinetic energy loss will be; therefore the lower the difference between η_{tt} and η_{ts} is. Values of $\alpha_5 \approx 0°$ are attained at higher rpms because of the lower Mach number at stator outlet/rotor inlet which in turn leads to a lower relative Mach at rotor outlet and thus can have lower absolute flow angles for similar pressure ratios at different speeds.

Table 10. Variation of relative flow angle at rotor inlet and absolute flow angle at rotor outlet at off-design conditions.

Π_{ts}	2	3	4	5	6	7
			β_4 [°]			
19,268 rpm	-1.07	43.97	49.88	51.10	51.28	51.25
21,409 rpm	-37.01	-44.19	40.92	44.21	45.29	45.65
23,550 rpm	-58.45	0.47	26.21	33.57	36.22	37.29
			α_5 [°]			
19,268 rpm	56.5	36.8	15.95	0.02	-10.64	-17.22
21,409 rpm	63.06	-27.85	26.09	9.56	-2.35	-10.27
23,550 rpm	69.81	48.48	34.49	18.65	5.87	-3.04

The total-to-static pressure ratio and rotational speed of the most and least efficient points are used in 3D CFD to visualise the flow behaviour for each case. Figure 12 shows the relative Mach number distribution for both cases at mid-span. One can note the absence of the shockwave and supersonic flows at the stator outlet channel for the conditions with high rpm which leads to much improved efficiency values ($\eta_{tt} = 82.40\%$ $\eta_{ts} = 72.39\%$). The right-hand-side figure (**b**) also shows that the inefficient turbine ($\eta_{tt} = 65.19\%$ $\eta_{ts} = 57.22\%$) exhibits loss features that include the presence of a vortex, incidence losses ($\beta_4 = 35.44°$) and suction side separation or flow reversal as explained in Section 3.3.2. Whereas, the left-hand-side figure (**a**) shows smooth flow acceleration and deceleration as would be ideal in a radial turbine case with proper rotor incidence of $\beta_4 = -12.93°$. The aim would be to design a turbine that coincides with such flow behaviour, however, operating conditions imposed by the cycle might not allow the elimination of supersonic flows. Investigation of the total pressure losses at the rotor exit hub-to-tip surface also emphasises the efficiencies at each off-design condition. A small fraction of tip leakage losses with a pressure loss region near the hub (25% P_0 loss near hub) corresponding to the better operating condition; whereas for the inefficient conditions, more pronounced shroud leakages with a proportionate value of 48% with respect to the areas of least pressure drop.

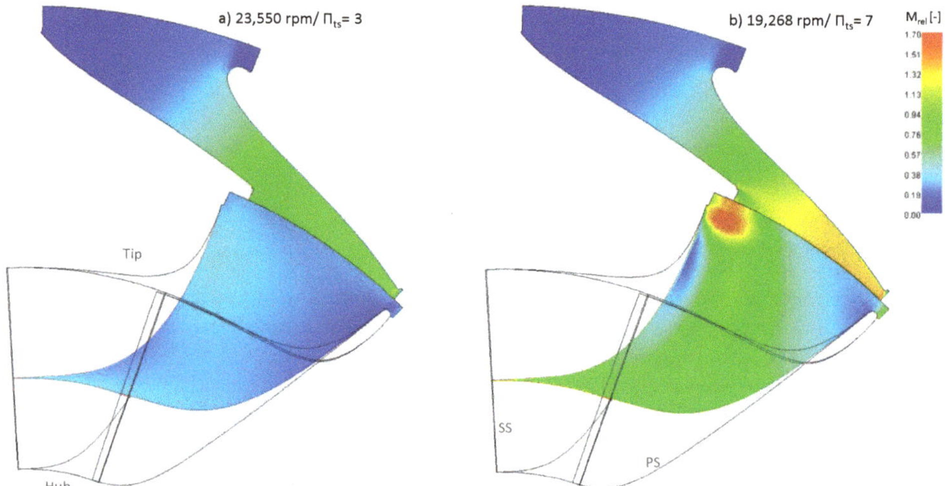

Figure 12. Off-design relative Mach number distribution at mid-span from 3D CFD simulations: (**a**) 23,550 rpm & $\Pi_{ts} = 3$, (**b**) 19,268 rpm & $\Pi_{ts} = 7$.

3.3.4. Isolated Rotor Analysis

An isolated rotor model is primarily developed for use in stress calculations but is also utilised for examination of the rotor performance without the inclusion of the stator effects. The boundary conditions are taken as dictated by the main project database of AxSTREAM with $P_{0_{in}} = 270$ bar, $T_{0_{in}} = 1173$ K at an absolute flow angle of 73.93° and $P_{out} = 54.0$ bar. Performance results of the rotor only model at design point are listed in Table 11. The computed total-to-static pressure drop is lower than the defined value of with the CFD value of $P_{0_{in}}$ being 256 bar. The mass flow rate is very close to the design 354.6 kg/s rate along with much better efficiencies compared to the full turbine cascade. The power output is around 14% lower than the 100 MW because of the lower pressure ratio (absence of stator means no static pressure drop there). The solution indicates that a large portion of losses arise in the stator-rotor passage and the incorrect flow conditions supplied by the nozzle.

Table 11. 3D CFD performance results for the isolated rotor.

Variable	Unit	Value
$\dot{m}$	kg/s	356.3
$\dot{W}$	MW	86.38
η_{tt}	%	87.80
η_{ts}	%	80.46
ψ	-	1.21
ϕ	-	0.423
R	-	0.406
Π_{ts}	-	4.75

The visualisation of several parameters is seen in Figure 13 where two blade passages are set side by side at mid-span. At the rotor leading edge, the flow experiences, as expected, an acceleration on the suction side and a slowing down on the pressure side with the corresponding pressure changes. There still exists a rise in entropy on the blade suction surface with the diffusion of the flow. However, relative velocity and total pressure distributions do not indicate full flow separation or reversal. In addition to that, the absence of the incidence loss region and the excessive nozzle acceleration allow for a more uniform flow distribution on the rotor blade.

The isolated rotor model gives better performance characteristics than the full turbine passage (stator and rotor). However, the power output is 86.38 MW which is lower than the target 100 MW range meaning that either a greater pressure drop is required, or a new design of an integrated nozzle is needed. In the former case, the higher pressure ratio can be obtained via higher velocities (increasing rpm or changing blade angles) however, it is important to maintain the speeds imposed by material limits.

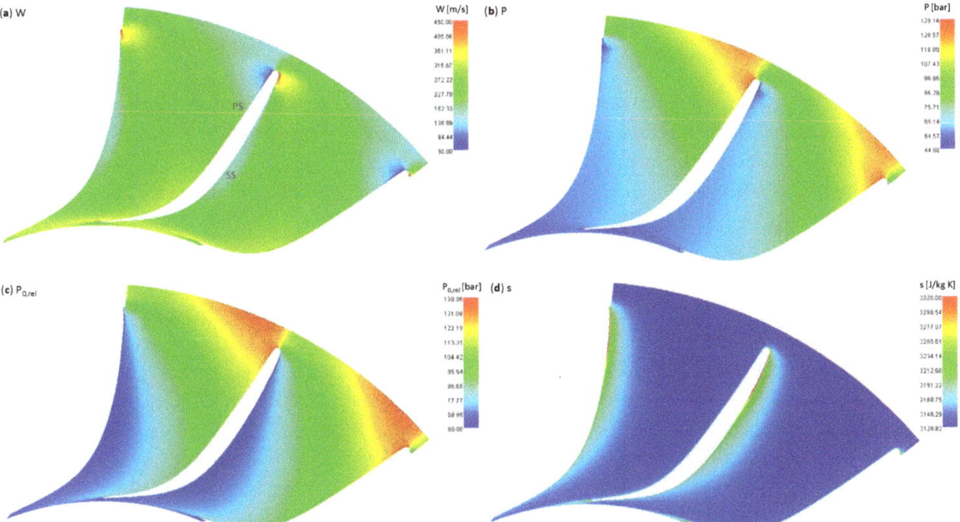

Figure 13. Mid-span parameters distribution for isolated rotor 3D CFD model.

A generic Nickel-based alloy (IN 718), typically employed in gas turbine applications is used in the stress simulations of the rotor disk. Although the known materials are not used for sufficient yield strengths at temperatures exceeding 760 °C [37], assumed properties are employed to investigate the range and areas of stress concentrations on a non-optimised rotor wheel design.

The static stress analysis performed signals a concern for the insufficient material strength at certain areas of the radial turbine. The aerodynamic part of the turbine itself (the designed rotor), even at higher loading than design point conditions, is mostly limited to stresses that are lower than the defined allowable von-Mises stress (400 MPa); the distribution of von Mises stresses is displayed using Figure 14. For better visualisation of the areas of high stress concentration, the contour limits of the displayed von Mises stress is set to the maximum 400 MPa; the grey regions are areas with stress over the permissible ceiling. The areas of stress concentration are defined as the disc root and the blade root both of which belong to the mechanical design aspect of the turbine. Regions of high stress reflect the need to revisit the mechanical design as part of the updated and optimised final design stage.

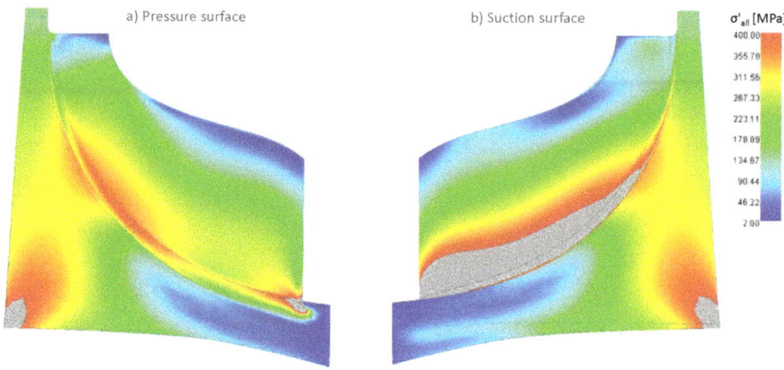

Figure 14. von-Mises stress distribution on rotor blade surfaces: (**a**) Pressure surface, (**b**) Suction surface.

4. Discussion on Enabling Technologies

Several technological barriers associated to the development of turbomachinery components for sCO$_2$ cycles have been identified in open literature [38]; a number of remarks can be made upon the outcomes presented in the paper based on the design of the radial turbine. The discussion to follow is rooted on information found in literature and on conversations with researchers working on related topics.

4.1. Materials Consideration

Material selection needs to be put at the forefront of the design considerations for the uncooled radial turbine for different reasons; firstly due to the creep and fatigue life, defined by material properties associated to the thermal and mechanical stresses and to the operating environment, and secondly for the synergistic effect of environment chemistry and parameters (temperature and pressure) on the material degradation. Materials have continuously evolved throughout the years to meet the high temperature operation market [39]. The proposed temperature of 900 °C sits on the limit of operating temperatures that do not require cooling for turbomachinery blades; nonetheless, temperature is not the only critical limitation for the proposed work.

The required strength of material at peak temperature with the mechanical stress resulting from the loads and rotational speed, and the high thermo-physical heat transfer properties of the fluid, might dictate the use of complex single-crystal superalloys if no cooling is employed; thermal barrier coatings can be explored to enable the use of less costly materials. Wright et al. [40] show that for the temperature limit of 900 °C the use of cast-alloy IN738 for turbine vanes and blades can be considered. There is a wide availability of materials, namely a range of steels, for the application in heat exchangers of state-of-the-art steam plants [41]. These cast alloys are more suitable for the turbomachinery casings, due to their high pressure capability, but not for use in the turbine blades where aero-mechanical stresses are significant.

The high inlet temperature of the turbine would dictate the use of superalloys usually employed in land based or aeroderivative gas turbines. Developments in metallurgy to reach single-crystal superalloys and coatings allow for a better creep deformation resistance in comparison to equiaxed crystal materials. Thus the extensive material database available for conventional gas and steam cycles is not appropriate for sCO$_2$ applications as where a narrow selection of the high pressure- high temperature materials is limited to expensive Ni-based alloys [42].

Erosion is reported in turbines using pure supercritical CO$_2$ so special attention needs to be given for erosion-resistant materials or coatings for the case of this cycle especially that the turbine operates with other contaminants coming from oxy-combustion [43]. Although short-term corrosion testing has been conducted by CSIRO for pure sCO_2 with temperatures and pressure scales up to 1000 °C and 250 bar respectively [41], results are limited because of the need to obtain more adequate long-term testing with actual conditions of impurities especially when H_2O is present in the working fluid. Within the same framework, increased corrosion has been reported with temperature rise; whereas the consequence of pressure change was found to be negligible. To mitigate the detrimental effect of internal oxidation and surface carburisation, the more expensive alloys with a high Chromium and Nickel content are necessitated; Austenitic steels have reduced corrosion resistance at high temperatures. Most oxidation and carburisation laboratory tests are also limited to ambient pressure environments with pure CO_2 fluids [40].

From the information stated above, one can deduce that the current array of materials still has major testing gaps, and is not suitable for the proposed operating conditions of the turbine (900 °C and 297 bar), unless cooling is incorporated. Collaborative efforts for all the material aspects is required for the actual adoption of the designed radial turbine.

4.2. Bearings

Bearings are of great design importance for the actual adaptation of the proposed radial turbine to provide support for the relative motion between the rotating parts. As the density of the fluid lowers from supercritical high density to subcritical conditions at the extreme pressure of the Allam cycle, a large axial force is generated on the turbine component; the axial thrust load issue is resolved using suitable thrust bearings [44]. Mechanical contact-type bearings such as rolling, sliding or flexing elements are not considered by researchers for sCO_2 applications. Non contacting, fluid-film hydrodynamic bearings are suggested by some [45,46] for their prevalence in high load/high speed industrial applications. A conclusion was obtained that oil-film bearings are not suitable for supercritical CO_2 utilisation because of the solvent-like characteristics of the sCO_2 fluid [47]. Recommendations include a specific hybrid hydrostatically-assisted hydrodynamic gas-foil bearing design, developed by Mechanical Solutions Inc. [48], that provides both radial and thrust support in a single combination component. The maximum temperature cited is up to 650 °C with a material coating which limits its applicability to the design in hand if no cooling is provided. Shaft size and weight will also limit the suitability since the inner diameter of the bearing geometry is 2 inches ($\approx$51 mm) which is smaller than the expected shaft diameter of 120 mm (calculated for mechanical design). Another possibility is rare earth element magnetic bearings which could levitate for 100 MW shaft scales and are developed by Electron Energy Corporation [49]. In any case, because of the operating temperature and pressure ranges of the designed turbine, bearings, and other components like gears, need to be isolated from the turbine itself via shaft end seals (as discussed in the next heading). Turbomachinery design manufacturers can provide answers regarding bearing design for the configuration of the designed turbine based on shaft sizes, but it seems that commercially-ready solutions are not yet available.

4.3. Shaft End Seals

Suitable shaft end seals for sCO_2 cycles can limit the leakage of the supercritical CO_2 fluid to minimise cycle efficiency penalties. They are required to separate the high pressure turbine from the rest of the ancillary components. Knowing that the temperature at turbine inlet is 900 °C, the seal at the end of the high-temperature/high-pressure side needs to be placed as far as possibly permissible from the turbine inlet without affecting rotordynamics. General Electric and the Southwest Research Institute [50] concluded that labyrinth seals are insufficient for providing the required low-leakage characteristics for the high-pressure fluid because the physical clearance is optimised for a single operating condition, and thus with the change of clearance gaps during different operation modes, leakage losses are aggravated. Non-contacting, dry gas lift-off face seals are suggested for high-temperature operation, with experimental testing by Sandia National Laboratories in collaboration with Flowserve Corporation at conditions of 700 °C, 4400 psi ($\approx$303 bar) and a shaft speed of 40,000 rpm [51]. In the case of the operating temperature of the modified NET Power cycle, Joule-Thomson or throttle cooling will need to be provided at that shaft end from compressor discharge to decrease the temperature to around 700 °C and allow safe and effective seal operation. Commercially-ready dry gas seals are limited to shaft sizes in the range of 37.5 mm for high rotational speeds (near 50,000 rpm) and for slower speeds, they can reach up to 10 times bigger shaft diameters [45]. Looking back at the conditions of the designed turbine in terms of temperature (900 °C), pressure (297 bar), rotational speed ($\approx$20,000 rpm) and shaft diameter ($\approx$120 mm), dry gas lift off seals are not a far measure from use for a 60–70 MW_e supercritical CO_2 radial turbine.

4.4. Electrical Configuration

The turbine shaft rotational speed is in the range of 20,000 rpm which would require an appropriate choice of a step-down gearbox to speeds of 1800 rpm or 3600 rpm for coupling the turbine to a synchronous generator or for the requirement of separating the unknown speed of the compressor which is not designed yet. Not only is the selection of the gearing configuration dependant on shaft speed but also on the power scale; McClung et al. [45] report that General Electric (GE) has gearing configuration for power ratings reaching 60 MW. The availability of a gearbox providing a speed reduction from 12,000 rpm to 3600 rpm for a 50 MW turbine scale is also disclosed by Bidkar et al. [52]. Geared turbine generator sets are widely available for steam turbines with power output up to 40 MW_e as reported by GE and Mitsubishi Hitachi [53,54]. Gearbox manufacturers can provide information on the applicability of having a speed reduction ratio of around 6 from near 20,000 rpm for a 60–70 MW_e turbine.

Besides the mentioned technologies which need to be developed for successful use of the radial turbine, there are some operational design considerations reported for sCO_2 turbines which require careful attention. Transient operation of the turbine might induce an overspeed risk and loss of electrical load due to the small size of the rotor. The necessitated temperature reduction near the turbine inlet, to safe margins tolerated by seals and bearings, incurs high thermal stresses in that region that needs addressing via further stress analysis.

5. Conclusions

This paper outlines the design procedure of a radial turbine for a cycle size of $\approx$100 MW starting with cycle modelling and moving towards preliminary design, performance analysis, CFD simulations, and mechanical assessment. Key findings include the identification of the areas within the turbine passage that require careful readjustment due to the presence of losses that diminish the performance. The final observation is that a radial turbine configuration is suitable for use in a utility-scale (100 MW) supercritical CO_2 power cycle at the pressure and temperature levels of 297 bar and 900 °C. The advantage of using a radial turbine is its compactness and its ability to endure a high pressure ratio in a single stage, about 5 for the designed turbine. The flow behaviour in the turbine design aligns well with results from other works on high fluid density radial turbines. However, the current design requires re-visitation to improve the overall turbine efficiency especially through optimising the design of the nozzle vanes through different stacking, or changing the number of blades to deliver better incidence to the rotor, and to reduce the exit Mach number, and thus the associated losses. The approach towards achieving the goal of employing a radial turbine in a mid-scale power plant requires further work. Future research should include, but is not limited to:

- Incorporation of the fluid mixture in the design processes to account for the influence on performance
- Optimisation of the blade profiles
- Additional stress analyses including modal, harmonic, and hot-to-cold simulations
- Experimental testing and validation
- Investigation of the relevant technologies such as seals, bearings, gearing
- Research on suitable material advancements

Author Contributions: T.E.S. performed the simulation and analysis, and drafted the manuscript. J.A.T. supervised the project, provided technical advice, and edited the manuscript. J.O. supervised the project, established research collaborations, and reviewed the manuscript. All authors have read and agreed to the published version of the manuscript.

Funding: This research received no external funding.

Acknowledgments: The authors would like to acknowledge the research group from Politecnico Di Milano that includes Prof. Paolo Chiesa, Prof. Emanuele Martelli and Roberto Scaccabarozzi for their collaboration on the cycle modelling part of the work. Also Prof. Giacomo Bruno Persico from the same institution for his input on the results of turbine design. The discussions with several researchers from the U.S. National Energy Technology Laboratory, Sandia National Laboratories, and Oak Ridge National Laboratories are highly valued. The authors appreciate the support from SoftInWay in matters relating to software.

Conflicts of Interest: The authors declare no conflict of interest.

Abbreviations

The following abbreviations are used in this manuscript:

Roman Symbols:

b	Passage width [mm]
c	Chord [mm]
M	Mach number [-]
$\dot{m}$	Mass flow rate [kg/s]
N	Number of blades [-]
n_s	Specific speed [-]
o	Throat [mm–m]
P	Pressure [bar]
q	Pitch [mm]
$\dot{Q}$	Volumetric flow rate [m^3/s]
R	Degree of reaction [-]
r	Radius [mm]
s	Entropy [J/kg K]
T	Temperature [K– °C]
t	Thickness [mm]
W	Relative velocity [m/s]
$\dot{W}$	Power [MW]
Z	Axial length [mm]

Greek Symbols:

α	Absolute flow angle [°]
β	Relative flow angle [°]
Δ	Difference [-]
η	Efficiency [%]
ν_s	Velocity ratio [-]
ω	Rotational speed [rad/s]
ϕ	Flow coefficient [-]
Π	Pressure ratio [-]
ψ	Stage loading coefficient [-]
ρ	Density [kg/m^3]
θ	Metal angle [°]

Subscripts:

0	Total value
2	Stator inlet station
3	Stator outlet station
4	Rotor inlet station
5	Rotor outlet station
e	Electric
id	Ideal

m   Meanline
max   Maximum
out   Outlet
R   Rotor
rel   Relative reference frame
S   Stator
th   Thermal
ts   Total-to-static
tt   Total-to-total

Acronyms/Abbreviations:

ASU	Air Separation Unit
CCS	Carbon Capture and Storage
CFD	Computational Fluid Dynamics
GE	General Electric
HTHE	High Temperature Heat Exchanger
IEA GHG	International Energy Agency GreenHouse Gas
LHV	Lower Heating Value
LTHE	Low Temperature Heat Exchanger
PS	Pressure Side
SS	Suction Side
TIP	Turbine Inlet Pressure
TIT	Turbine Inlet Temperature

References

1. Ahn, Y.; Lee, J.I. Study of Various Brayton Cycle Designs for Small Modular Sodium-Cooled Fast Reactor. *Nucl. Eng. Des.* **2014**, *276*, 128–141. [CrossRef]
2. Dahlquist, A.; Genrup, M. Aerodynamic Turbine Design for an Oxy-Fuel Combined Cycle. In Proceedings of the ASME Turbo Expo 2016: Turbomachinery Technical Conference and Exposition (GT 2016), Seoul, Korea, 13–17 June 2016; Volume 2.
3. Dahlquist, A.; Genrup, M.; Sjoedin, M.; Jonshagen, K. Optimization of an Oxyfuel Combined Cycle Regarding Performance and Complexity Level. In Proceedings of the ASME Turbo Expo 2013: Turbine Technical Conference and Exposition, San Antonio, TX, USA, 3–7 June 2013.
4. Zhang, N.; Lior, N. Two Novel Oxy-Fuel Power Cycles Integrated with Natural Gas Reforming and CO_2 Capture. *Energy* **2008**, *33*, 340–351. [CrossRef]
5. Olumayegun, O.; Wang, M.; Kelsall, G. Closed-Cycle Gas Turbine for Power Generation: A State-of-the-Art Review. *Fuel* **2016**, *180*, 694–717. [CrossRef]
6. Kato, Y.; Nitawaki, T.; Muto, Y. Medium Temperature Carbon Dioxide Gas Turbine Reactor. *Nucl. Eng. Des.* **2004**, *230*, 195–207. [CrossRef]
7. Turchi, C.; Ma, Z.; Neises, T.W.; Wagner, M.J. Thermodynamic Study of Advanced Supercritical Carbon Dioxide Power Cycles for Concentrating Solar Power Systems. *J. Sol. Energy Eng.* **2013**, *135*, 041007. [CrossRef]
8. Allam, R.; Palmer, M.; Brown, G.; Fetvedt, J.; Freed, D.; Nomoto, H.; Itoh, M.; Okita, N.; Jones, C. High Efficiency and Low Cost of Electricity Generation from Fossil Fuels while Eliminating Atmospheric Emissions, Including Carbon Dioxide. *Energy Procedia* **2013**, *37*, 1135–1149. [CrossRef]
9. IEAGHG. *Oxy-Combustion Turbine Power Plants*; IEAGHG: Cheltenham, UK, 2015.
10. Scaccabarozzi, R.; Gatti, M.; Martelli, E. Thermodynamic Analysis and Numerical Optimization of the NET Power Oxy-Combustion Cycle. *Appl. Energy* **2016**, *178*, 505–526. [CrossRef]
11. Allam, R.; Fetvedt, J.; Forrest, B.; Freed, D. The Oxy-Fuel, Supercritical CO_2 Allam Cycle: New Cycle Developments to Produce Even Lower-Cost Electricity From Fossil Fuels Without Atmospheric Emissions. In Proceedings of the ASME Turbo Expo 2014: Turbine Technical Conference and Exposition. American Society of Mechanical Engineers, Düsseldorf, Germany, 16–20 June 2014.
12. Allam, R.; Martin, S.; Forrest, B.; Fetvedt, J.; Lu, X.; Freed, D.; Brown, G.; Sasaki, T.; Itoh, M.; Manning, J. Demonstration of the Allam Cycle: An Update on the Development Status of a High Efficiency Supercritical Carbon Dioxide Power Process Employing Full Carbon Capture. *Energy Procedia* **2017**, *114*, 5948–5966. [CrossRef]

13. Aspen Technology Inc. *Aspen Plus User Guide 10.2*; Aspen Technology Inc.: Bedford, MA, USA, 2000.

14. White, M.; Sayma, A. A Preliminary Comparison of Different Turbine Architectures for a 100kW Supercritical CO_2 Rankine Sycle Turbine. In Proceedings of the 6th International Supercritical CO_2 Power Cycles Symposium, Pittsburgh, PA, USA, 27–29 March 2018.

15. Holaind, N.; Bianchi, G.; De Miol, M.; Saravi, S.; Tassou, S.; Leroux, A.; Jouhara, H. Design of Radial Turbomachinery for Supercritical CO_2 Systems using Theoretical and Numerical CFD Methodologies. *Energy Procedia* **2017**, *123*, 313–320. [CrossRef]

16. Zhang, H.; Zhao, H.; Deng, Q.; Feng, Z. Aerothermodynamic Design and Numerical Investigation of Supercritical Carbon Dioxide Turbine. In Proceedings of the ASME Turbo Expo 2015: Turbine Technical Conference and Exposition, Montreal, QC, Canada, 15–19 June 2015.

17. Wang, Y.; Li, J.; Zhang, D.; Xie, Y. Numerical Investigation on Aerodynamic Performance of SCO_2 and Air Radial-Inflow Turbines with Different Solidity Structures. *Appl. Sci.* **2020**, *10*, 2087. [CrossRef]

18. Aungier, R. *Turbine Aerodynamics Axial-Flow and Radial-Inflow Turbine Design and Analysis*; ASME, Three Park Avenue: New York, NY, USA, 2006.

19. Rahbar, K.; Mahmoud, S.; Al-Dadah, R. Mean-Line Modeling and CFD Analysis of a Miniature Radial Turbine for Distributed Power Generation Systems. *Int. J. Low Carbon Technol.* **2016**, *11*, 157–168. [CrossRef]

20. Moroz, L.; Govorushchenko, Y.; Pagur, P.; Grebennik, K.; Kutrieb, W.; Kutrieb, M. Integrated Environment for Gas Turbine Preliminary Design. In *Osaka International Gas Turbine Congress*; Citeseer: Osaka, Japan, 2011; pp. 13–18.

21. Shah, S.; Chaudhri, G.; Kulshreshtha, D.; Channiwalla, S. Radial Inflow Gas Turbine Flow Path Design. *Int. J. Eng. Res. Dev.* **2013**, *5*, 41–45.

22. Aungier, R. A Fast, Accurate Real Gas Equation of State for Fluid Dynamic Analysis Applications. *J. Fluids Eng.* **1995**, *117*, 277–281. [CrossRef]

23. Meijboom, L. Development of a Turbine Concept for Supercritical CO_2 Power Cycles. Master's Thesis, Delft University of Technology, Delft, The Netherlands, 2017.

24. Dong, B.; Xu, G.; Luo, X.; Zhuang, L.; Quan, Y. Analysis of the Supercritical Organic Rankine Cycle and the Radial Turbine Design for High Temperature Applications. *Appl. Therm. Eng.* **2017**, *123*, 1523–1530. [CrossRef]

25. Pecnik, R.; Rinaldi, E.; Colonna, P. Computational Fluid Dynamics of a Radial Compressor Operating with Supercritical CO_2. *J. Eng. Gas Turbines Power* **2012**, *134*, 122301. [CrossRef]

26. Wei, Z. Meanline Analysis of Radial Inflow Turbines at Design and Off-Design Conditions. Master's Thesis, Carleton University, Ottawa, ON, Canada, 2014.

27. Takagi, K.; Muto, Y.; Ishizuka, T.; Kikura, H.; Aritomi, M. Research on Flow Characteristics of Supercritical CO_2 Axial Compressor Blades by CFD Analysis. In Proceedings of the 17th International Conference on Nuclear Engineering ICONE17, Brussels, Belgium, 12–16 July 2009; Volume 4, pp. 565–572.

28. Horlock, J.; Torbidoni, L. Calculations of cooled turbine efficiency. *J. Eng. Gas Turbines Power* **2008**, *130*, 011703. [CrossRef]

29. Wilcock, R.; Young, J.; Horlock, J. The Effect of Turbine Blade Cooling on the Cycle Efficiency of Gas Turbine Power Cycles. *J. Eng. Gas Turbines Power* **2005**, *127*, 109–120. [CrossRef]

30. Fleming, D.; Holschuh, T.; Conboy, T.; Rochau, G.; Fuller, R. Scaling Considerations for a Multi-Megawatt Class Supercritical CO_2 Brayton Cycle and Path Forward for Commercialization. In Proceedings of the ASME Turbo Expo 2012: Turbine Technical Conference and Exposition, Copenhagen, Denmark, 11–15 June 2012; pp. 953–960.

31. Dixon, S.; Hall, C. *Fluid Mechanics and Thermodynamics of Turbomachinery*; Butterworth-Heinemann: Oxford, UK, 2010.

32. Watch, M. Gas Turbine Market Size 2018- Global Analysis, Comprehensive Research Study, Development Status, Business Growth, Competitive Landscape, Future Plans and Trends by Forecast 2023. 2019. https://www.marketwatch.com/press-release/ (accessed on 18 December 2019).

33. Cerdoun, M.; Ghenaiet, A. CFD Analyses of a Radial Inflow Turbine. In Proceedings of the 8th International Conference on Compressors and Their Systems, London, UK, 9–10 September 2013; Woodhead Publishing: Sawston, UK, 2013; pp. 635–647.

34. Monteiro, V.; Zaparoli, E.; de Andrade, C.; de Lima, R. Numerical Simulation of Performance of an Axial Turbine First Stage. *J. Aerosp. Technol. Manag.* **2012**, *4*, 175–184. [CrossRef]

35. Sauret, E.; Gu, Y. 3D CFD Simulations of a Candidate R143a Radial-Inflow Turbine for Geothermal Power Applications. In Proceedings of the ASME 2014 Power Conference: Turbine Technical Conference and Exposition, Düsseldorf, Germany, 16–20 June 2014.

36. Baines, N. *Introduction to Radial Turbine Technology*; VKI Radial Turbines: Sint-Genesius-Rode, Belgium, 1992.

37. Metals, H.T. Inconel 718 Technical Data. 2015. https://www.hightempmetals.com/techdata/hitempInconel718data.php (accessed on 18 December 2019).

38. Brun, K.; Friedman, P.; Dennis, R. *Fundamentals and Applications of Supercritical Carbon Dioxide (sCO$_2$) Based Power Cycles*; Woodhead Publishing: Sawston, UK, 2017.

39. Barua, S. High-Temperature Hot Corrosion of Gas Turbine Materials. Master's Thesis, Cranfield University, Cranfield, UK, 2018.

40. Wright, I.; Pint, B.; Shingledecker, J.; Thimsen, D. Materials Considerations for Supercritical CO$_2$ Turbine Cycles. In Proceedings of the ASME Turbo Expo 2013: Turbine Technical Conference and Exposition, San Antonio, TX, USA, 3–7 June 2013.

41. Subbaraman, G.; Kung, S.; Saari, H. Materials for Supercritical CO$_2$ Applications. In Proceedings of the 6th International Supercritical CO$_2$ Power Cycles Symposium, Pittsburgh, PA, USA, 27–29 March 2018.

42. Essilfie-Conduah, N. Operating Conditions and Materials Selection for Gas-Fired Supercritical CO$_2$. Master's Thesis, Cranfield University, Cranfield, UK, 2019.

43. Finn, J.; He, X.; Apte, S. Erosion in Components of Supercritical CO$_2$ Power Cycles. In *Utsr Project Review Meeting*; National Energy Technology Laboratory: Pittsburgh, PA, USA, 2018.

44. Cho, J.; Choi, M.; Baik, Y.; Lee, G.; Ra, H.; Kim, B.; Kim, M. Development of the Turbomachinery for the Supercritical Carbon Dioxide Power Cycle. *Int. J. Energy Res.* **2016**, *40*, 587–599. [CrossRef]

45. McClung, A.; Smith, N.; Allison, T.; Tom, B. Practical Considerations for the Conceptual Design of an sCO$_2$ Cycle. In Proceedings of the 6th International Supercritical CO$_2$ Power Cycles Symposium, Pittsburgh, PA, USA, 27–29 March 2018.

46. Allison, T.; Wilkes, J.; Brun, K.; Moore, J. Turbomachinery Overview for Supercritical CO$_2$ Power Cycles. In Proceedings of the 46th Turbomachinery and 33rd Pump Symposia, Houston, TX, USA, 12–14 December USA, 2017; Southwest Research Institute: San Antonio, TX, USA, 2017.

47. Fleming, D. *Private Communication-Telephone*; Sandia National Laboratories: Albuquerque, NM, USA, 2019.

48. Chapman, P. Advanced Gas Foil Bearing Design for Supercritical CO$_2$ Power Cycles. In Proceedings of the 5th International Supercritical CO$_2$ Power Cycles Symposium, San Antonio, TX, USA, 28–31 March 2016.

49. Liu, J.; Dent, P.; Palazzolo, A. *Novel Ultra High Temperature Magnetic Bearings for Space Vehicle Systems*; Technical Report; Electron Energy Corporation: Landisville, PA, USA, 2018.

50. Bidkar, R.; Sevincer, E.; Wang, J.; Thatte, A.; Mann, A.; Peter, M.; Musgrove, G.; Allison, T.; Moore, J. Low-Leakage Shaft-End Seals for Utility-Scale Supercritical CO$_2$ Turboexpanders. *J. Eng. Gas Turbines Power* **2016**, *139*, 022503. [CrossRef]

51. Fleming, D. *Experience with < 1MW Supercritical Power Generation Technology Development*; Technical Report; Sandia National Laboratories: Albuquerque, NM, USA, 2017.

52. Bidkar, R.; Mann, A.; Singh, R.; Sevincer, E.; Cich, S.; Day, M.; Kulhanek, C.; Thatte, A.; Peter, A.; Hofer, D.; Moore, J. Conceptual Designs of 50 MWe and 450 MWe Supercritical CO$_2$ Turbomachinery Trains for Power Generation from Coal. Part 1: Cycle and Turbine. In Proceedings of the 5th International Symposium—Supercritical CO$_2$ Power Cycles, San Antonio, TX, USA, 28–31 March 2016.

53. Systems, G.P. Steam Turbines for Industrial Applications. Available online: https://www.ge.com/content/dam/gepower-pgdp/global/en_US/documents/technical/ger/ger-3706d-steam-turbines-industrial-applications.pdf (accessed on 1 March 2020).

54. Systems, M.H.P. Geared Turbines. Available online: https://www.mhps.com/products/steamturbines/lineup/industrial/deceleration/ (accessed on 18 March 2020).

Article

Numerical Investigation on Aerodynamic Performance of SCO$_2$ and Air Radial-Inflow Turbines with Different Solidity Structures

Yuqi Wang [1], Jinxing Li [2], Di Zhang [1],* and Yonghui Xie [2]

[1] MOE Key Laboratory of Thermo-Fluid Science and Engineering, School of Energy and Power Engineering, Xi'an Jiaotong University, Xi'an 710049, China; wyq_xjtu@163.com

[2] Shaanxi Engineering Laboratory of Turbomachinery and Power Equipment, School of Energy and Power Engineering, Xi'an Jiaotong University, Xi'an 710049, China; xing837439166@stu.xjtu.edu.cn (J.L.); yhxie@mail.xjtu.edu.cn (Y.X.)

* Correspondence: zhang_di@mail.xjtu.edu.cn; Tel.: +86-29-8266-6559

Received: 24 February 2020; Accepted: 9 March 2020; Published: 19 March 2020

Featured Application: This work can provide the design reference of rotor solidity for SCO$_2$ and air radial-inflow turbines as well as a new splitter structure to improve the performance of low solidity case.

Abstract: Supercritical carbon dioxide (SCO$_2$) is of great use in miniature power systems. It obtains the characteristics of high density and low viscosity, which makes it possible to build a compact structure for turbomachinery. For a turbine design, an important issue is to figure out the appropriate solidity of the rotor. The objective of this research is to present the aerodynamic performance and provide the design reference for SCO$_2$ and air radial-inflow turbines considering different solidity structures. For the low solidity case of SCO$_2$ turbine, new splitter structures are proposed to improve its performance. The automatic design and simulation process are established by batch modes in MATLAB. The numerical investigation is based on a 3D viscous compressible N-S equation and the actual fluid property of SCO$_2$ and air. The distributions of flow parameters are first presented. Rotor blade load and aerodynamic force are then thoroughly analyzed and the aerodynamic performances of all cases are obtained. The SCO$_2$ turbine has larger power capacity and higher efficiency while the performance of the air turbine is less affected by rotor solidity. For both SCO$_2$ and air, small solidity can cause the unsatisfactory flow condition at the inlet and the shroud section of the rotor, while large solidity results in the aerodynamic loss at the trailing edge of rotor blade and the hub of rotor outlet. A suction side offset splitter can greatly improve the performance of the low solidity SCO$_2$ turbine.

Keywords: radial-inflow turbine; supercritical carbon dioxide; air; rotor solidity; aerodynamic performance

1. Introduction

In recent years, the study of supercritical carbon dioxide (SCO$_2$) Brayton cycle and its components has attracted lots of attention. Various heat sources including solar power [1–5], nuclear power [6,7] and waste-heat utilization [8] are employed. As the key component in a Brayton cycle, the design and characteristics of turbomachinery deserve to be investigated in depth. SCO$_2$ has a critical point around room temperature (7.38 MPa, 304.25 K). As a working fluid in a power cycle, SCO$_2$ has a large number of advantages. First of all, it is environmentally friendly and the critical point is easy to reach. Hence, it is safe and cheap in industrial applications. Additionally, the high density and low viscosity of SCO$_2$ can result in the high efficiency and compact mechanical structure of turbines and compressors in the power system [9–11]. The rapid changes of density near the critical point can reduce the compressor work and lead to a higher heat transfer coefficient in heat regenerators and precoolers [12]. Finally, there is little phase change in a SCO_2 power cycle. Thus, it needs fewer valves and no condenser, which results in a concise cycle.

To realize this proposed technology, Sandia National Laboratory has developed several SCO_2 power cycles including various power class [13]. The radial turbine and compressor test rigs were established to investigate the key techniques. However, the outcomes of efficiency and rotation speed were far from the expectations. Hence, the design method and flow mechanism need to be further considered before the commercialization of this technique. A traditional design process of the turbomachinery is time-consuming. It needs several engineers to conduct multiple numerical calculation to obtain the case of best performance. Hence, it is vital to establish an automatic process in order to save the manpower and accelerate the design.

The flow conditions are rather complex in turbomachinery, which results in the difficulty of turbine design. In the past years, the research of design, influence of key parameters and optimization are widely covered [14–17]. Solidity, as a design parameter, plays an important role in turbine design. It was proposed by Zweifel [18] in 1945, in which he estimated the optimum solidity for turbines with large angular deflection. However, Horlock [19] pointed out that the prediction of Zweifel's optimum solidity correlation was limited to the outlet flow angle of the blade. In the past investigations, the influence of solidity on axial wind turbine performance occupied the vast majority. Chen et al. [20] investigated the effects of flanged diffusers on rotor performance of small wind turbines with different rotor solidities and wind speeds. Mohamed [21] studied the effect of the turbine solidity and the usage of hybrid system between drag and lift types on small wind turbine performance numerically and experimentally. Eboibi et al. [22] experimentally investigated the influence of solidity on the performance and flow field aerodynamics of vertical axis wind turbines at low Reynolds numbers. Gao et al. [23] thoroughly analyzed the effects of rotor solidity and leakage flow on the unsteady flow in axial turbine.

Some researches focus on the solidity of stator vane in radial-inflow turbines. Simpson et al. [24] conducted the numerical and experimental study of the performance effects of varying vaneless space and vane solidity in radial turbine stators. Pereiras et al. [25] concentrated on the influence of the guide vane solidity on the performance of a radial impulse turbine with pitch-controlled guide vanes. Dong et al. [26] evaluated the effects of outlet blade angle, solidity, blade height, expansion ratio, and surface roughness on the stator velocity coefficient.

To sum up, the stator solidity has been considered in radial-inflow turbines while the choice of rotor solidity is rarely investigated, especially its impact in a SCO_2 turbine. As the working fluid approaches the critical point at the outlet of a SCO_2 turbine and the value of solidity tends to decrease along the flow direction in a radial-inflow rotor, the concentration of rotor solidity is needed. We establish an automatic design process by batch modes in MATLAB 2019b of MathWorks (Natick, MA, USA) with an accurate numerical simulation method of radial-inflow turbines. Several solidity structures are considered and new splitter structures are proposed and analyzed to improve the performance of low solidity case. The distributions of flow parameters are first presented. Rotor blade load and aerodynamic force are then thoroughly analyzed and the aerodynamic performances are obtained. The design reference of rotor solidity for SCO_2 and air turbines are provided.

2. Modeling

2.1. Establishment of Automatic Design Process

To accelerate the design process, an automatic design and calculation process of the radial-inflow turbine is firstly established by calling batch modes in MATLAB. The flow chart is shown in Figure 1. The initial geometric model of the turbine is obtained based on design parameters, such as inlet pressure and temperature, outlet pressure, output power, rotation speed, etc. Then, the rotor solidity and working fluid used in the calculation are determined. The macro files of batch mode are called by MATLAB to execute the process in Workbench BladeGen, Turbogrid, CFX, and CFD-Post. The modeling, discretization, numerical simulation and post-processing are then completed. Finally, the program repeats this procedure to obtain the results of the required different cases.

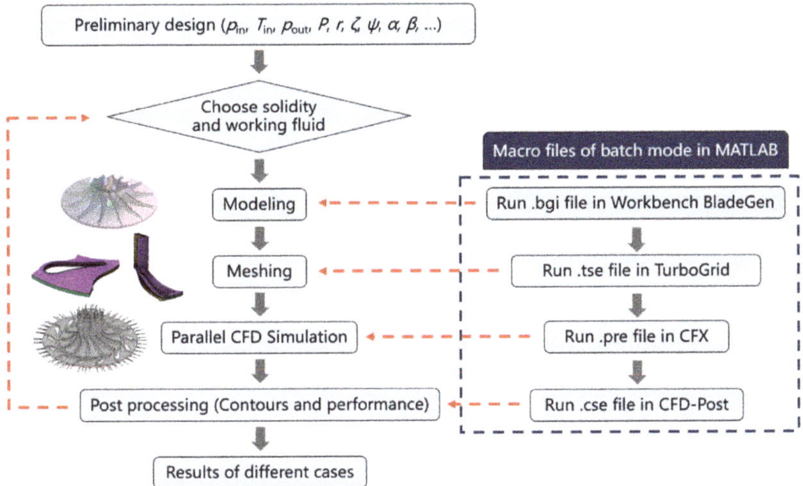

Figure 1. Flow chart of design and simulation.

The thermodynamic design is based on the conservation of momentum, mass and energy equations. The fluid properties of SCO_2 and air are obtained from NIST REFPROP. The basic principal for the thermodynamic design is to ensure the highest efficiency at a given rotation speed [27]. Due to space limits, the following formulas are some of the calculations in the thermodynamic design program. The influence of friction loss at the wheel back and the flow loss in the volute are ignored. The loss producing by nozzle and impeller are considered. Hence, the isentropic efficiency can be estimated as Equation (1):

$$\eta_{is} = 1 - (1 - \alpha)\left(1 - \zeta^2\right) - \left(\frac{\omega_{out}}{C_{is}}\right)^2\left(\frac{1}{\psi^2} - 1\right) \tag{1}$$

where α is the degree of reaction. ζ and ψ are the velocity coefficient of nozzle and impeller respectively, which represents the loss in nozzle and impeller. We adopt 0.96 for ζ and 0.84 for ψ. ω_{out} and C_{is} are, respectively, the relative velocity at turbine outlet and the isentropic ideal velocity.

To gain the best isentropic efficiency in the thermodynamic design, we can correspondingly acquire the degree of reaction, the velocity ratio and etc. For example, the degree of reaction is chosen as 0.47 and the velocity ratio is 0.67. After the determination of design parameters, we can calculate the geometry parameters. For example, the impeller diameter is acquired by Equation (2):

$$D = \frac{60v_{in}}{\pi r} \tag{2}$$

where r is the rotating speed, which is 50,000 rpm in this case. v_{in} is the linear velocity at the inlet of the impeller.

The blade height at the turbine inlet can be calculated as Equation (3):

$$l_{in} = \frac{\dot{m}}{\pi D \rho_{in} \omega_{in}} \tag{3}$$

where $\dot{m}$ is the mass flow rate, ρ_{in} and ω_{in} are, respectively, the density and the relative velocity at the rotor inlet.

The area of turbine outlet is calculated in Equation (4):

$$A_{\text{out}} = \frac{\dot{m}}{\rho_{\text{out}}\omega_{\text{out}}\sin\beta_{\text{out}}} \tag{4}$$

where β_{out} is the flow angle at turbine outlet, ρ_{out} and ω_{out} are, respectively, the density and the relative velocity at rotor outlet. With the empirical coefficients regarding to impeller diameter, the average outlet diameter and axial length are determined. Then we can obtain the geometric parameters of meridian plane for the rotor.

The geometric parameters are calculated by estimating the velocity at both inlet and outlet of the stator and the rotor. To compare the difference of air and SCO$_2$, we firstly conduct the thermodynamic design using SCO$_2$ and then, estimate the working condition with air to conduct the numerical simulation with the same radial-inflow turbine model.

2.2. Research Model

The working fluid first flows into the nozzle flow passage. In the nozzle, the pressure gradually declines and accordingly, the velocity increases. Figure 2 presents the geometry model of nozzle. A straight blade with uniform geometry angles from hub to shroud direction is adopted due to the low blade height. The geometry of nozzle is identical in all conditions with different solidity structures. The red part represents high temperature and pressure while the blue part stands for lower temperature and pressure. Detailed geometric parameters, i.e., the diameter of inlet and outlet, the blade height, and the geometry angle of the inlet and outlet are given in Figure 2.

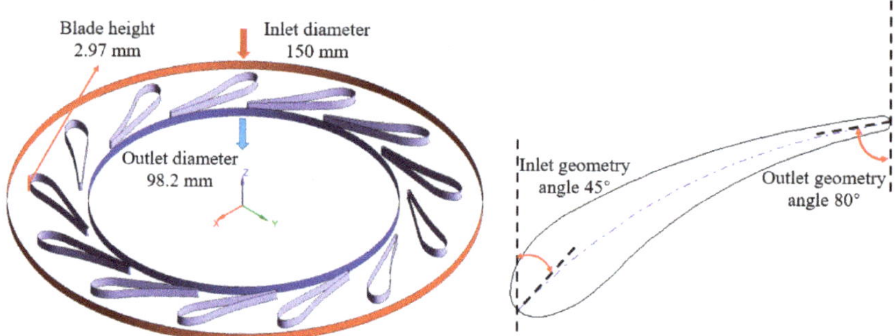

Figure 2. Geometry model of nozzle.

After the process in the nozzle, the working fluid enters from a radial direction into the impeller. In this process, SCO$_2$ expands in the impeller and exits from axial direction. The backward bent vane is adopted to guarantee the designed velocity triangle and a shrouded impeller is used to promote the aerodynamic efficiency. Figure 3 shows the geometry model of impeller, taking blade number 12, as an example. After the calculation of impeller diameter, blade height of inlet and outlet, external diameter of outlet and axial length, the arc to connect each point is determined by design experience. In Figure 3, the red arrow stands for high temperature and pressure while the blue arrow represents relatively low temperature and pressure. Brennen [28] gave the definition of solidity for radial pumps. Likewise, we define the solidity with cord length and pitch in a radial-inflow turbine. These parameters are obtained in Workbench BladeGen while modelling. In this research, the rotor solidity is defined by c/s, where c is the cord length of the rotor blade and s is the trailing edge pitch at the outlet of the impeller, as shown in Figure 3.

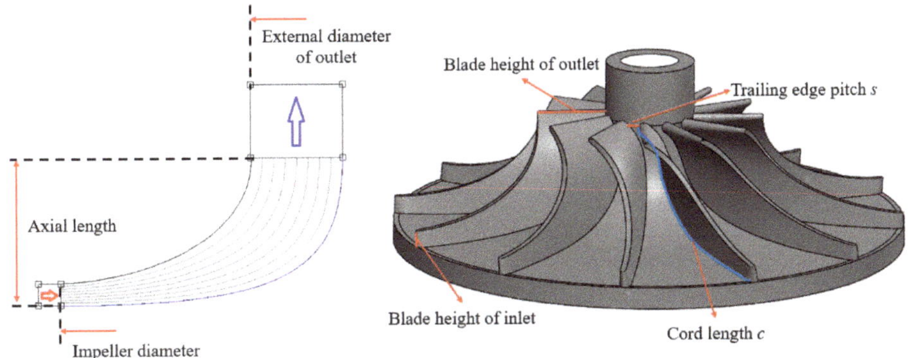

Figure 3. Geometry model of the impeller.

The concrete geometric parameters of impeller are shown in Table 1. Generally, the number of nozzle blades and the number of impeller blades should be relatively prime for the consideration of reducing the exciting force of turbomachinery. Hence, while the nozzle blade number is designed as 13, we choose five kinds of rotor solidity, which are realized by changing the rotor blade numbers. They are even from 8 to 16. The large range of rotor solidity can present its effect on aerodynamic performance.

Table 1. Geometric parameters of the impeller.

Parameter	A	B	C	D	E
Blade number	8	10	12	14	16
Impeller diameter/mm			92.2		
External diameter of outlet/mm			40.6		
Axial length/mm			20		
Blade height of inlet/mm			2.97		
Blade height of outlet/mm			24.6		
Inlet geometry angle/°			90		
Outlet geometry angle/°			27		
Chord length (c)/mm			50		
Pitch at trailing edge (s)/mm	6.28	5.03	4.19	3.59	3.14
Solidity (c/s)	7.96	9.95	11.94	13.93	15.92

High solidity usually results in large weight and axial thrust of the wheel. In some application environments such as aerospace and warships, the weight of the wheel is limited. The overlarge axial thrust and the trans-critical problem of the high solidity case also need consideration in the turbine design. Hence, for the low solidity case of SCO_2 turbine mentioned above, we try to establish a new structure to improve the flow condition at the leading edge of the rotor blade. A splitter, which is normally used in centrifugal compressors, is introduced here. We establish a pressure side (ps) offset splitter and a suction side (ss) offset splitter to compare with the normal one, as shown in Figure 4. The offset value is 20% of the local pitch, that is, the distance between the splitter blade the closest main blade is 30% of the local pitch.

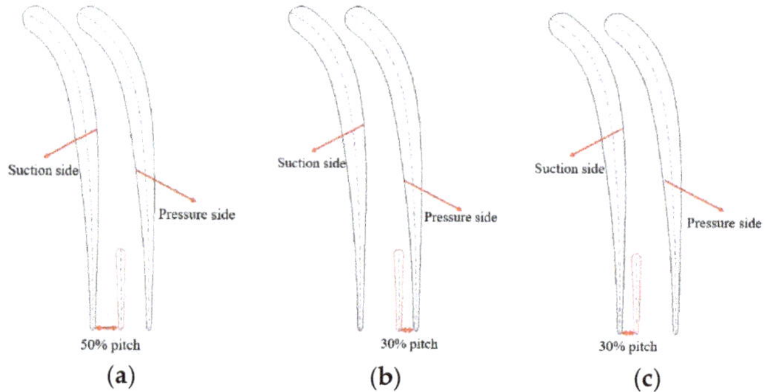

Figure 4. Splitter structures for the lowest solidity case: (**a**) normal; (**b**) ps offset; (**c**) ss offset.

3. Methodology

A 3D viscous compressible CFD simulation is carried out in commercially-available software CFX 18.2. In this research, the computing resource of an i7-3770 CPU and eight threads is used. In order to obtain the specific characteristics of SCO_2 and air radial-inflow turbines with different solidity structures, an accurate numerical simulation model needs to be established. The following sections introduce the validation of turbulence model, working fluids as well as the discretization method used in numerical calculations.

3.1. Validation of Computational Method

In CFD simulation, the conservation equation of mass, energy and momentum are solved by Reynolds-Averaged Navier–Stokes (RANS) method. In this investigation, a compressible and steady turbulent flow is simulated. The equations can be expressed as below:

$$\frac{\partial(\rho u_i)}{\partial x_j} = 0 \tag{5}$$

$$\frac{\partial\left(\rho u_i u_j\right)}{\partial x_j} = -\frac{\partial p}{\partial x_i} + \frac{\partial\left(\mu \frac{\partial x_i}{\partial x_j} - \rho\overline{u_i' u_j'}\right)}{\partial x_j} + S_i \tag{6}$$

$$\frac{\partial(\rho u_i T)}{\partial x_j} = -\frac{\partial\left(\Gamma \frac{\partial T}{\partial x_j} - \rho\overline{u_i' T'}\right)}{\partial x_j} + S \tag{7}$$

where u stands for velocity components, x is displacement term, S_i and S are, respectively, the source term of momentum and inner heat. To handle the governing equations of turbulent flow, the RANS method is adopted. Hence, a turbulence model is needed for the calculation of Reynolds stress.

To discretize the governing equations, an element-based finite volume method is adopted. TurboGrid 18.2 is employed to generate the hexahedral meshes in all fluid domains, which correspond to lower truncation error. To obtain the best accuracy and the influence of the wake flow, a full cycle of the impeller and the nozzle is adopted in the investigation. Figure 5 shows the mesh of the fluid domain including the partially enlarged view of the leading edge of the nozzle, the rotor-stator interface and the trailing edge of the rotor. O-type mesh is applied around the rotor blades and nozzle blades. The mesh of the boundary layers has been densely generated to adapt the turbulence model. Concretely, the discretization in the boundary layer area ensures the averaged solver y^+ is set around 1as recommended in the CFX User Guide.

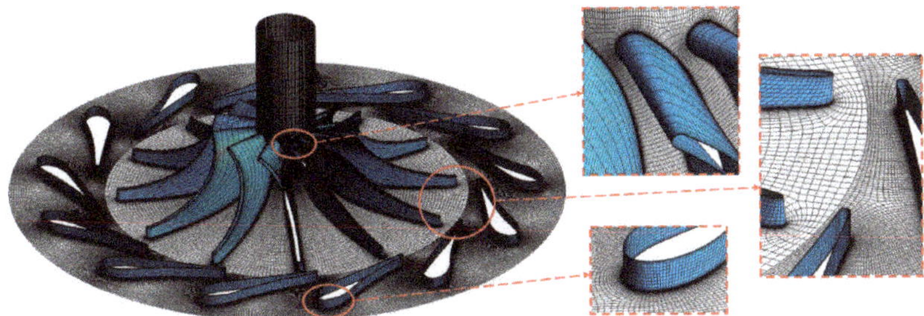

Figure 5. Mesh of the computational field.

For the determination of near-wall velocities, the automatic wall function is chosen. Total energy equation including the viscous work term is used for energy conservation. A high-resolution advection scheme of CFX is used. The boundary conditions employed in this research are listed in Table 2. We use total pressure and temperature as inlet boundary conditions. To simulate the inlet volute, the flow angle at the nozzle inlet is set as 45°. The static pressure is used for outlet boundary condition while the rotation speed is set as 50,000 rpm for the rotor. A frozen-rotor method is adopted to handle the interface between stator and rotor. An automatic domain initialization method is used.

Table 2. Boundary conditions of the calculated cases.

Working Fluid	Total Pressure of Inlet/MPa	Total Temperature of Inlet/K	Static Pressure of Outlet/MPa	Rotation Speed/rpm
SCO$_2$	15	550	8	50,000
Air	0.5	550	0.3	50,000

The computation is considered converged when RMS residuals are below 10^{-5}. In the meantime, we evaluate the mass flow rate of the turbine inlet and outlet. If they are equal and slightly fluctuate with each iteration step, the case reaches convergence.

In order to validate the method of CFD simulation, we choose a similar radial-inflow turbine working with air from Deng et al. [29], which possesses experimental results. A CFD simulation with Deng's model and boundary conditions under design point is conducted. Four common turbulence models for turbomachinery simulation including k-ε, RNG k-ε, k-ω, and SST k-ω are adopted and the validation results are given in Figure 6. It can be seen from Figure 6 that compared with the results in [29], the mass flow rate and efficiency calculated by k-ε models are higher. Compared with the experiment, we adopt a shrouded impeller in the design. The leakage flow derived from tip clearance is not taken into account. Besides, due to the simplification of the research model, the leakage flow between the nozzle and impeller is not considered in the numerical simulation. These cause the simulated mass flow rate higher than Deng's experiment. Comparing four turbulence models, SST k-ω turbulence model has the smallest error. The relative errors of the mass flow and efficiency are, respectively, 3.9% and −0.12%. Therefore, we use the SST k-ω turbulence model in the CFD analysis. The turbulence model was first proposed by Menter [30] and it was based on the k-ε and k-ω turbulence models. This model can accurately simulate the wall region and does not acquire high mesh quality as in the k-ω turbulence model.

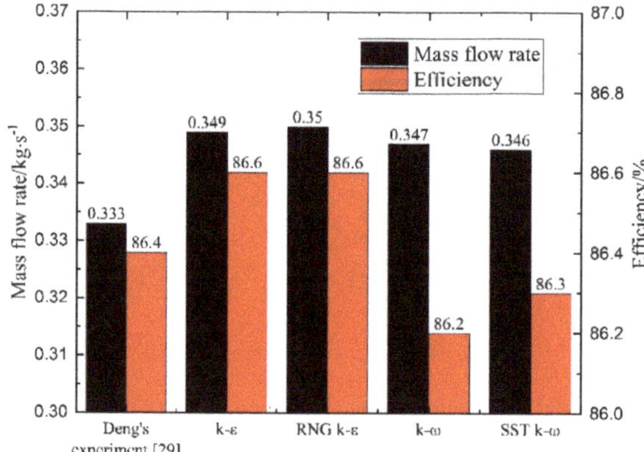

Figure 6. Validation of different turbulence models.

To guarantee the accuracy of the numerical simulation, a mesh independence test has been conducted under the condition of solidity 11.94. Grid convergence index (GCI) measurement is employed, which was proposed by Roache [31] to provide an objective asymptotic approach to quantification of uncertainty of grid convergence. It is based on the theory of generalized Richardson Extrapolation and involves the comparison of discrete solutions at different grid resolutions. The method of GCI is commonly employed in the grid refinement study [32].

To calculate the GCI, three mesh sizes are used with a constant refinement ratio $r = 1.2$. The value of F_s is chosen 1.25 as suggested by Roache [31]. The results are presented in Table 3. The convergence condition is evaluated by the convergence ratio $R = (f_3 - f_2)/(f_2 - f_1)$, which is 0.625 considering mass flow rate and 0.333 considering output power. Hence, it corresponds to monotonic convergence in this case. The small value of GCI indicates the reduced dependency of results on mesh size. When the total element number is 3,354,000, GCI_{12} for mass flow rate and output power are below 1%. Hence, the fine mesh scheme (f_1) is accurate to conduct the numerical simulation of all cases.

Table 3. Result of grid convergence index.

Mesh	Total Element Number/10^4	Mass Flow Rate/kg·s^{-1}	GCI/%	Output Power/kW	GCI/%
Coarse(f_3)	113.2	6.581		343.03	
Medium(f_2)	193.6	6.549	$GCI_{23} = 1.02$ $GCI_{12} = 0.64$	342.16	$GCI_{23} = 0.16$ $GCI_{12} = 0.05$
Fine(f_1)	335.4	6.529		341.87	

3.2. Validation of Fluid Property

For the working fluid air, the fixed composition mixture is set up in CFX 18.2 with the state equation of Peng Robinson. This option is commonly used in CFX to model the real gas properties for gas density, enthalpy, etc.

In the simulation of a SCO$_2$ turbine, it is of great importance to precisely obtain the fluid property of SCO$_2$. For the simulated condition, the state of working fluid lies in the supercritical region. To simulate the fluid property precisely, we use the real gas property (RGP) format table to implement the sharply variable density and specific heat capacity near the critical point. The tabulated pressure and temperature region are, respectively, 7 MPa to 16 MPa and 300 K to 600 K in order to simulate the fluctuated state parameter during the converging process.

The parameters for the RGP table are obtained from NIST REFPROP, which is widely referred to as a fluid property database [33]. MATLAB is used to generate the RGP format file with a predefined temperature and pressure range. Figure 7 gives the density variation with the temperature and pressure range for the SCO$_2$ turbine.

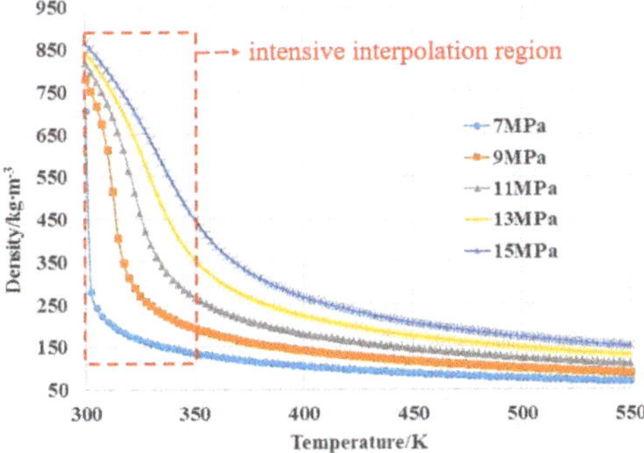

Figure 7. Density variation with temperature and pressure of SCO$_2$.

As shown in Figure 7, the density of the working fluid changes sharply around the critical point. The CFX solver calculates the SCO$_2$ property by bilinear interpolation between parameter points. Hence, we choose to densely generate the RGP table near the critical point for the balance of accuracy and computation time. The region is shown in the dashed box in Figure 7. For the temperature range, we adopt a gradient of 2 K for 300–350 K while for 350–600 K, we use a gradient of 5 K. For the pressure range, we utilize 0.1 MPa as a gradient between 7–8 MPa while 0.2 MPa as a gradient is employed between 8–16 MPa. Compared with 2 K for all temperature range and 0.1 MPa for all pressure range, the computation time is effectively reduced by 20.4%.

The accuracy of RGP format table using in ANSYS CFX has been validated by Odabaee et al. [34] and Ameli et al. [35]. We conduct a validation in the selected range of this research. Three points are chosen to validate the precision of RGP file, which are, respectively, the inlet, outlet and the interface between stator and rotor. We consider four parameters to test the relative error between CFX calculation and data from NIST, which are density, enthalpy, C_p and dynamic viscosity. As shown in Table 4, the maximum relative error is 0.009%. Hence, the RGP table is sufficiently accurate for the simulation of SCO$_2$ working fluid.

Table 4. Validation of SCO$_2$ fluid property.

Parameter	Density/kg·m^{-3}	Enthalpy/kJ·kg^{-1}	C_p/J·kg^{-1}·K^{-1}	Dynamic Viscosity/Pa·s
Location	Inlet			
CFX	173.19	648.10	1243.97	2.6859×10^{-5}
NIST	173.19	648.10	1244.00	2.6859×10^{-5}
Relative error/%	0	0	0.002	0
Location	Interface between stator and rotor			
CFX	144.15	630.05	1211.87	2.5113×10^{-5}
NIST	144.14	630.05	1211.80	2.5111×10^{-5}
Relative error/%	0.006	0	0.006	0.008
Location	Outlet			
CFX	106.35	601.88	1159.11	2.2715×10^{-5}
NIST	106.35	601.88	1159.10	2.2717×10^{-5}
Relative error/%	0	0	0.001	0.009

4. Results and Discussion

After the automatic design calculation by batch modes in MATLAB, we obtain the aerodynamic performance of cases with different solidity structures and working fluid. In this chapter, we first present the flow parameters to understand the flow mechanism of different solidity structures and working fluid. Then the rotor blade load and aerodynamic force are compared and analyzed. Finally, the aerodynamic performance is presented to estimate the best solidity and the effectiveness of new splitter structures are tested for low solidity case.

4.1. Distributions of Flow Parameters

4.1.1. Pressure Distribution

The pressure distribution reveals the power capacity of the turbine. In this section, the pressure distribution of the SCO_2 and air turbine are presented and compared. Figure 8 shows the pressure distribution at the mid-span of the SCO_2 radial-inflow turbine. In a SCO_2 radial-inflow turbine, the rotor solidity gradually increases along the flow direction as the pitch at leading edge is much wider than that of trailing edge. As shown in Figure 8, when the rotor solidity is the minimum, the suction side of the rotor blade exists an apparent low-pressure region and the pressure distribution is non-uniform. With the rotor solidity increasing, the low-pressure region tends to decrease and the working capacity for each blade decreases. The diminishing space of rotor flow passage restricts the flow of SCO_2 fluid. Hence, the expansion of working fluid tends to be light in (d) and (e). Condition (c) has a most uniform pressure distribution and the working capacity of blades is satisfying. In condition (a) and (b), the large space at the inlet of the rotor causes the disorder of the pressure distribution.

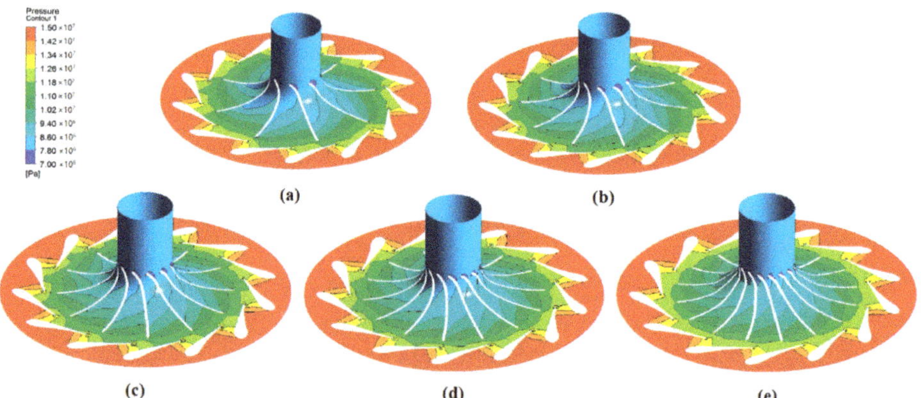

Figure 8. Pressure distribution at mid-span of SCO_2: (**a**) solidity 7.96; (**b**) solidity 9.95; (**c**) solidity 11.94; (**d**) solidity 13.93; (**e**) solidity 15.92.

As is known to all, when the temperature or pressure is around critical point, the physical property of SCO_2 can be sharply variable, especially for density and specific heat capacity. In the turbomachinery with working fluid SCO_2, it is a priority problem to minimize the region where SCO_2 transforms from a supercritical state to a subcritical state to obtain stable operation. We term the transformation from the supercritical state to the subcritical state as a trans-critical phenomenon. Hence, Figure 9 presents the trans-critical phenomenon at the trailing edge. It is a vital position in radial-inflow turbine to analyze.

As Figure 9 presents, the trans-critical phenomenon is obvious at the trailing edge of the SCO_2 radial-inflow turbine. In condition (a) and (b), the region area is rather small and it concentrates on the tip of the rotor blade. As the rotor solidity increases, the pitch at trailing edge decreases. Hence, the flow area of SCO_2 reduces, which distinctly extends the area of trans-critical region. As we can see, in condition (c), the minimum pressure along the fluid domain is below 6.5 MPa. With the further decrease of flow area, trans-critical region also appears at the root of rotor blade. In (d) and (e), there exists a local constriction at the trailing edge of rotor blade, which causes an entire trans-critical region at the pressure side of trailing edge.

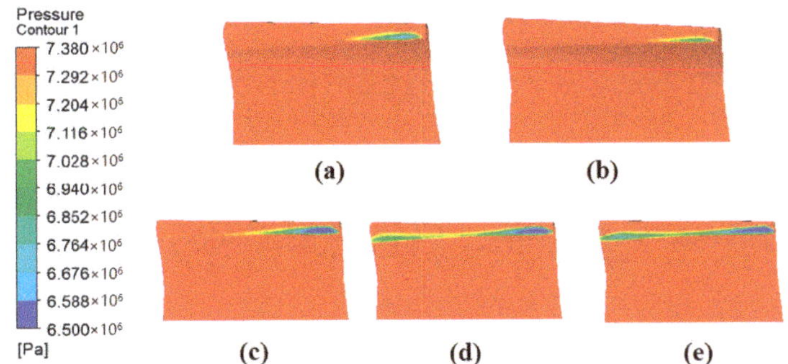

Figure 9. Trans-critical phenomenon at the trailing edge: (**a**) solidity 7.96; (**b**) solidity 9.95; (**c**) solidity 11.94; (**d**) solidity 13.93; (**e**) solidity 15.92.

For comparison, Figure 10 shows the pressure distribution at the mid-span with air as working fluid. Likewise, when the rotor solidity is the minimum, the pressure gradient is not uniform in the middle of the flow passage. A low-pressure region presents at the suction side of rotor blade in Figure 10a,b. In (c) and (d), the pressure distribution is uniform while in (e), the pressure difference between suction side and pressure side of the blade decreases. It is worth mentioning that in an air turbine, the low-pressure region at the suction side of rotor trailing edge vanishes. There is no trans-critical phenomenon in the air turbine.

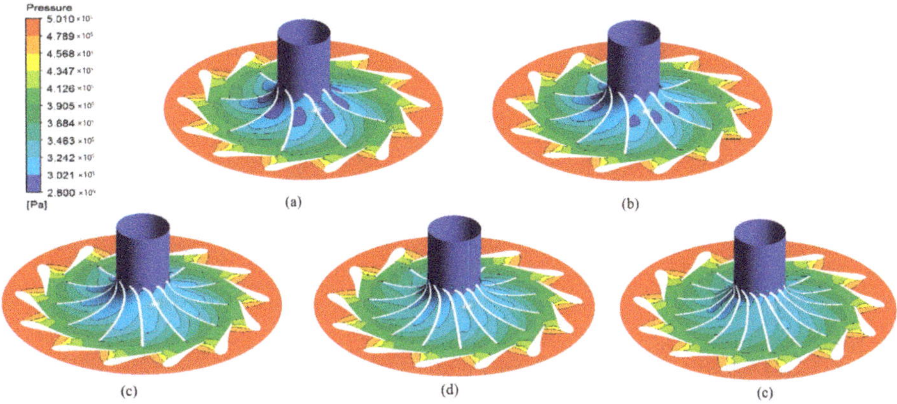

Figure 10. Pressure distribution at mid-span of air: (**a**) solidity 7.96; (**b**) solidity 9.95; (**c**) solidity 11.94; (**d**) solidity 13.93; (**e**) solidity 15.92.

The averaged pressure along spanwise location at turbine outlet are presented in Figure 11, where (a) stands for SCO_2 and (b) is air. When the value of spanwise location is 1, it corresponds to the shroud while value zero corresponds to the hub. As Figure 11a presents, the distribution of the averaged pressure along spanwise location at turbine outlet is quite different with the change of rotor solidity in a SCO_2 turbine. Especially when the solidity is 15.92, the power capacity at the different cross-section of the blade varies a great deal. The local constriction at the hub results in more pressure loss and it corresponds to the lowest pressure. While at the shroud, the expansion of working fluid is incomplete, which results in the highest pressure. In general, the cross-section of the hub has the highest power ability. Comparing five kinds of solidity, the pressure along spanwise tends to be more uniform as the solidity decreases. This is because higher solidity greatly disturbs the flow between blades, which is not beneficial for the safe operation of the turbine. In Figure 11b, the tendency is similar in an air turbine except that the pressure along spanwise is less affected by solidity. The working fluid air has no sharp change of fluid property at the outlet when the solidity is high.

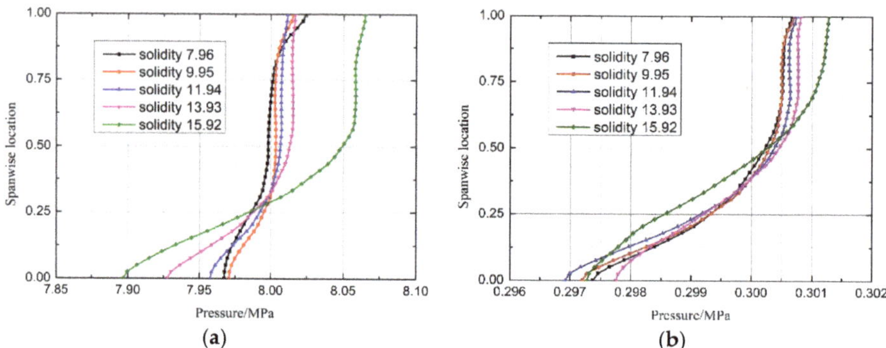

Figure 11. Averaged pressure along spanwise location at turbine outlet: (**a**) SCO_2; (**b**) air.

4.1.2. Entropy Distribution

The static entropy distribution can reflect the aerodynamic loss in the flow passage. For different rotor solidity, the difference of aerodynamic loss mainly concentrates on the outlet. In this section, the static entropy at the outlet of SCO_2 and air are presented and compared.

The averaged static entropy along spanwise location at turbine outlet are presented in Figure 12. Spanwise location 1 and 0 correspond to the shroud and hub, respectively. The green plane is the contour location. It is acknowledged that higher entropy corresponds to higher aerodynamic loss. For both SCO_2 and air, the aerodynamic loss mainly concentrates on the hub and shroud area due to the boundary layer effect of the impeller. The distribution is quite different with the change of rotor solidity. At the shroud, the highest static entropy appears due to the relatively vacant flow passage when solidity is 7.96 in the SCO_2 turbine. For air, it locates at solidity 9.95, which means the decrease of solidity less affects the aerodynamic loss at the shroud. While at the hub, the highest static entropy presents when solidity is 15.92 on account of the limited flow area for both two working fluids.

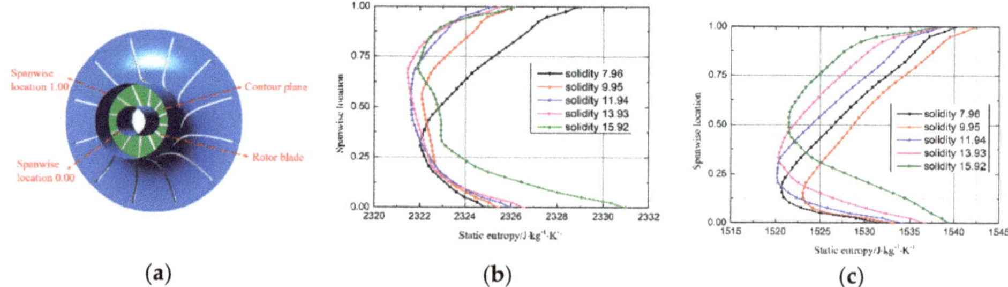

Figure 12. Averaged static entropy along spanwise location at turbine outlet: (**a**) location of graphical results; (**b**) SCO_2; (**c**) air.

Figure 13 shows the entropy distribution at outlet with surface streamlines of the SCO_2 radial-inflow turbine. When the rotor solidity is the minimum, the local entropy production in the flow passage near the shroud is apparent. Due to the low solidity, the disordered flow near the shroud results in several vortexes, thus causing local entropy concentration. With the rotor solidity increases, the vortexes gradually disappear and the increase of local entropy near the shroud completely vanishes in condition (d), i.e., solidity 13.93. The location of entropy concentration changes to the flow passage near the hub. When rotor solidity is the minimum, the flow near the shroud is disordered while solidity is the highest, the diminishing space of the rotor flow passage restricts the flow of SCO_2 fluid. These reasons cause the local enlargement of entropy as well as aerodynamic loss.

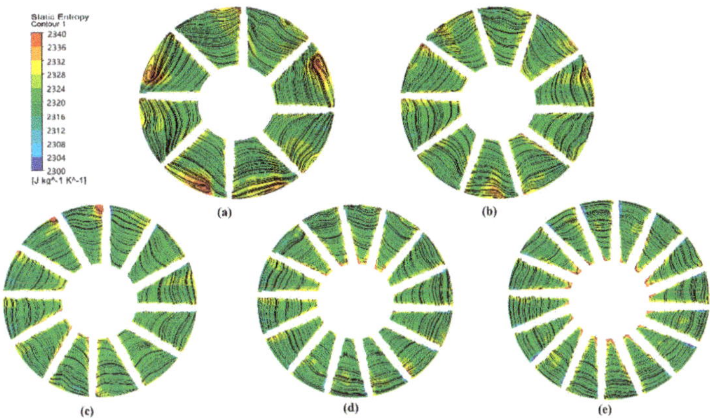

Figure 13. Entropy distribution at outlet with surface streamlines of SCO_2: (**a**) solidity 7.96; (**b**) solidity 9.95; (**c**) solidity 11.94; (**d**) solidity 13.93; (**e**) solidity 15.92.

Comparing to Figures 13 and 14 shows the entropy distribution at outlet with surface streamlines of the radial-inflow turbine working with air. The static entropy of the air turbine is apparently lower. In Figure 14a,b, the local vortex at the shroud causes local entropy production while in the SCO_2 turbine, the vortex nearly disappears when solidity is 9.95. With the solidity further increasing, the entropy production still concentrates on the shroud side even when the solidity becomes the highest. Hence, in the air turbine, the diminishing space of rotor flow passage does not cause the local enlargement of aerodynamic loss at the hub.

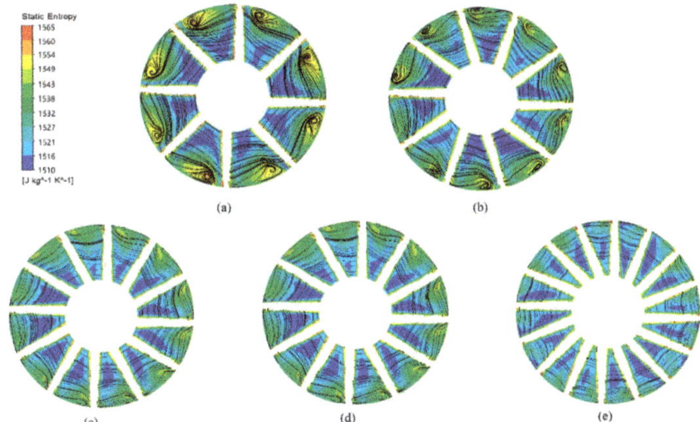

Figure 14. Entropy distribution at outlet with surface streamlines of air: (**a**) solidity 7.96; (**b**) solidity 9.95; (**c**) solidity 11.94; (**d**) solidity 13.93; (**e**) solidity 15.92.

4.2. Rotor Blade Load and Aerodynamic Force

The power capability of the radial-inflow turbine depends on rotor blade load and it can be further studied by rotor blade pressure distributions. Figure 15 presents the pressure distribution at the mid-span and its partially enlarged view, i.e., the rotor blade load along the streamwise location of different solidity and working fluid. As Figure 15a indicates, when solidity is higher than 11.94, the pressure below critical point appears. The trans-critical phenomenon arises at the trailing edge of the blade. In both Figure 15a,b, when solidity is 7.96, the pressure difference between suction side and pressure side is the largest. Hence, the blade can obtain the highest power capacity when the solidity is the lowest. Additionally, the capacity of the blade concentrates on the front and middle of the streamwise location, especially for solidity 15.92, which has little power output at the latter half of the blade. This phenomenon derives from the diminishing pitch of the blade along the flow direction. Moreover, for both SCO_2 and air, the solidity mainly affects the pressure distribution on the suction side and the pressure on the pressure side is of minor variation.

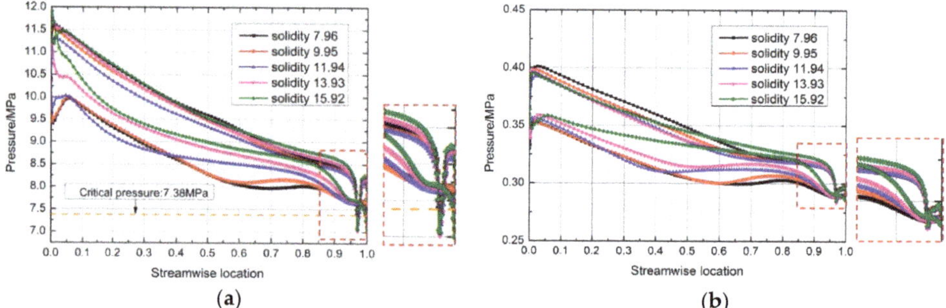

Figure 15. Rotor blade load along streamwise location of different solidity: (**a**) SCO_2; (**b**) air.

The axial force of a single rotor blade with different solidity and working fluid are presented in Figure 16. The direction of the force is shown by the red arrow. The axial force of the rotor is not beneficial to the safe operation and requires high-level balance method of axial thrust as well as a reliable bearing. From Figure 16, we can conclude that with the solidity increasing, in a SCO_2 radial-inflow turbine, the axial force of a single rotor blade increases. However, with the working fluid air, the axial force first increases and when the solidity is greater than 13.93, it slightly decreases. In general, the axial thrust of the rotor increases considering the rotor blade number. The effect of rotor solidity on axial force is greater for SCO_2 turbine than it is for air turbine. The high density of SCO_2 working fluid causes much higher axial thrust. With the augmented axial force, the safe operation of the turbine is affected and the load of the bearing enlarges.

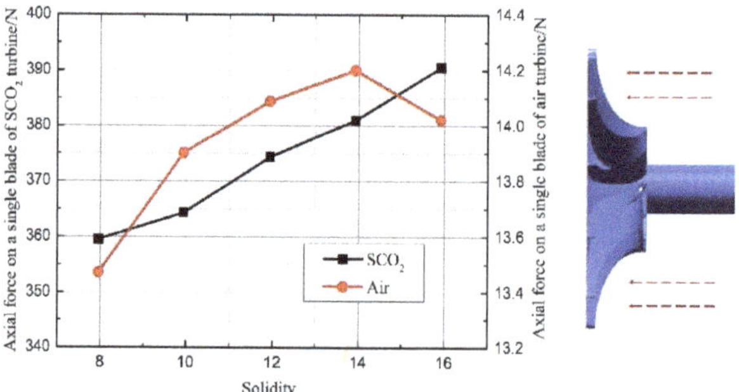

Figure 16. Axial force of a single rotor blade with different solidity and working fluid.

4.3. Aerodynamic Performance

Output power, isentropic efficiency and mass flow rate are vital parameters in the assessment of the turbine aerodynamic performance. The isentropic efficiency discussed in this research is defined as Equation (8):

$$\eta_{is} = \frac{2T_Z \times 2\pi \times r}{60 \times \left(\dot{m}_{in} + \dot{m}_{out}\right)\Delta h_{is}} \tag{8}$$

where T stands for torque, r represents the rotation speed, h is enthalpy and $\dot{m}$ is mass flow rate. As for the subscripts, is stands for isentropic, in and out represent the inlet and outlet of the turbine respectively, z corresponds to axial direction. The mass flow rate of inlet and outlet are slightly different due to the random error in the numerical simulation. To be precise, we adopt the average to calculate the efficiency.

High solidity usually results in large weight and axial thrust of the wheel. The working fluid SCO_2 has more special application scenarios (such as aerospace and warships) than air due to its small size. In these cases, the weight of the wheel is limited. The problems of overlarge axial thrust and trans-critical phenomenon need to be solved. Hence, we concentrate on the improvement of SCO_2 turbine. Three new splitters structures with and without offsetting are introduced. All solidity structures are calculated and compared in this section to test the effectiveness in the SCO_2 radial-inflow turbine. Figure 17 gives the variation of output power and isentropic efficiency with different solidity.

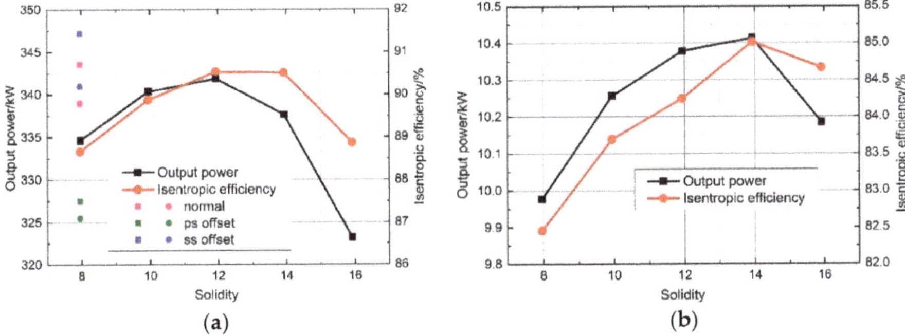

Figure 17. Output power and isentropic efficiency of different solidity and splitters: (**a**) SCO_2; (**b**) air.

For both SCO_2 and air, the output power of the turbine first increases and then decreases with the augment of solidity. For SCO_2, the maximum output power is 341.87 kW and it is realized when solidity is 11.94. While in Figure 17b, the maximum output power is 10.41 kW and it happens when solidity is 13.93 for the air turbine. Comparing (a) and (b), the radial-inflow turbine working with air has apparently lower power capacity, which results from a lower power density. With the augment of solidity, the output power descends more rapidly after the maximum in SCO_2 turbine than that in air turbine. The decline of output power is owing to the blocking effect of turbine outlet when solidity reaches a certain extent. Obviously, it has a greater influence in a SCO_2 turbine due to the fluid property similar to liquid.

As for the isentropic efficiency, the tendency is similar for SCO_2 and air. In general, the isentropic efficiency first increases and then decreases with the augment of solidity and the efficiency of SCO_2 turbine is higher in this power scale. In a SCO_2 turbine, when solidity is 11.94, the isentropic efficiency reaches the maximum 90.54%. When solidity is 13.93, the efficiency is 90.52%. Although the highest solidity corresponds to the lowest output power, its efficiency is acceptable due to the lowest mass flow rate. The solidity 7.96 has the minimum efficiency owing to the unsatisfactory flow condition at the inlet of impeller, which is caused by the large pitch at leading edge. In the air turbine, the highest efficiency is 85.02% when solidity is 13.93. Likewise, the isentropic efficiency is the lowest 82.46% when solidity is 7.96. The efficiency declines slower than that in SCO_2 turbine after the solidity reaches the maximum. The blocking effect is smaller and the large pitch at the leading edge affects the efficiency in a large scale.

The ss offset splitter can increase the output power and efficiency to the greatest extent. The power is increased to 347.22 kW, and the efficiency is increased to 90.18%. The normal splitter can improve the performance to a small extent while the ps offset splitter is ineffective. Overall, in the preliminary design of a SCO_2 radial-inflow turbine, the best solidity is around 12 and for an air turbine, it is around 14. If the weight, axial thrust and trans-critical problem is concerned, arranging the ss offset splitter is a choice to improve the flow condition and the aerodynamic performance of the low solidity case.

5. Conclusions

In this research, an automatic design and simulation process of the radial-inflow turbine is established in MATLAB. The design reference of rotor solidity for both SCO_2 and air are provided. A new splitter structure to improve the aerodynamic performance of low solidity case is proposed. Concretely:

1. For both SCO_2 and air, with the increase of rotor solidity, the working capacity for each blade declines. The cross section of hub has the highest power ability. Comparing five kinds of solidity, the pressure along spanwise tends to be more uniform as the solidity decreases. Small rotor solidity can result in the chaotic flow at the inlet of impeller, which further leads to the nonuniform pressure distribution. Large solidity disturbs the flow at the outlet and the blocking flow increases the trans-critical area in SCO_2 turbine. Comparing two working fluids, the air turbine has no trans-critical area and the low-pressure region at the suction side of the rotor blade trailing edge vanishes.

2. The aerodynamic loss of the SCO_2 and air turbine mainly concentrates on the hub and shroud area due to the boundary layer effect of the impeller. For SCO_2, when solidity is 7.96, the highest static entropy locates at the shroud due to the vacant flow passage. When solidity is 15.92, the highest static entropy locates at the hub on account of the limited flow area. For air, the increase of solidity less affects the aerodynamic loss and the entropy production mainly concentrates on the shroud.

3. For both SCO_2 and air, the solidity mainly affects the pressure distribution on the suction side and the pressure on the pressure side is of minor variation. In the SCO_2 turbine, large solidity corresponds to lower power capacity and greater trans-critical area of each blade. The high density of SCO_2 working fluid causes much higher axial thrust than air.

4. For both SCO_2 and air, the output power and isentropic efficiency of the turbine first increases and then decreases with the augment of solidity. In the SCO_2 turbine, the isentropic efficiency reaches the maximum 90.54% and the maximum output power is 341.87 kW when solidity is 11.94. While in the air turbine, the maximum isentropic efficiency is 85.02% and the maximum output power is 10.41 kW when solidity is 13.93. In the preliminary design of a SCO_2 radial-inflow turbine, the best solidity is around 12 and for an air turbine, it is around 14. Arranging the ss offset splitter can improve the flow condition and the aerodynamic performance of the low solidity case.

Author Contributions: Conceptualization: Y.W. and D.Z.; Methodology: D.Z.; Software: Y.X.; Validation: Y.W. and J.L.; Formal analysis: Y.W.; Investigation: Y.W. and J.L.; Resources: Y.X.; Data curation: D.Z.; Writing—original draft preparation: Y.W.; Writing—review and editing: J.L.; Visualization: J.L.; Supervision: D.Z.; Project administration: Y.X.; funding acquisition: Y.X. All authors have read and agreed to the published version of the manuscript.

Funding: This research was funded by 111 project, grant number B16038.

Conflicts of Interest: The authors declare no conflict of interest.

References

1. Garg, P.; Kumar, P.; Srinivasan, K. Supercritical carbon dioxide Brayton cycle for concentrated solar power. *J. Supercrit. Fluids* **2013**, *76*, 54–60. [CrossRef]
2. Chen, M.-F.; Yamaguchi, H.; Zhang, X.-W.; Niu, X.-D. Performance analyses of a particularly designed turbine for a supercritical CO_2-based solar Rankine cycle system. *Int. J. Energy Res.* **2015**, *39*, 1819–1827. [CrossRef]
3. Osorio, J.D.; Hovsapian, R.; Ordonez, J.C. Dynamic analysis of concentrated solar supercritical CO_2-based power generation closed-loop cycle. *Appl. Therm. Eng.* **2016**, *93*, 920–934. [CrossRef]
4. Chacartegui, R.; de Escalona, J.M.M.; Sanchez, D.; Monje, B.; Sanchez, T. Alternative cycles based on carbon dioxide for central receiver solar power plants. *Appl. Therm. Eng.* **2011**, *31*, 872–879. [CrossRef]
5. Valdes, M.; Abbas, R.; Rovira, A.; Martin-Aragon, J. Thermal efficiency of direct, inverse and sCO2 gas turbine cycles intended for small power plants. *Energy* **2016**, *100*, 66–72. [CrossRef]
6. Dostal, V.; Hejzlar, P.; Driscoll, M.J. High-performance supercritical carbon dioxide cycle for next-generation nuclear reactors. *Nucl. Technol.* **2006**, *154*, 265–282. [CrossRef]
7. Ahn, Y.; Bae, S.J.; Kim, M.; Cho, S.K.; Baik, S.; Lee, J.I.; Cha, J.E. Review of supercritical CO_2 power cycle technology and current status of research and development. *Nucl. Eng. Technol.* **2015**, *47*, 647–661. [CrossRef]
8. Mecheri, M.; Le Moullec, Y. Supercritical CO_2 Brayton cycles for coal-fired power plants. *Energy* **2016**, *103*, 758–771. [CrossRef]
9. Monje, B.; Sanchez, D.; Savill, M.; Pilidis, P.; Sanchez, T.; ASME. *A Design Strategy for Supercritical CO_2 Compressors*; America Society Mechanical Engineers: New York, NY, USA, 2014.
10. Zhang, H.; Zhao, H.; Deng, Q.; Feng, Z. *Aerothermodynamic Design and Numerical Investigation of Supercritical Carbon Dioxide Turbine*; America Society Mechanical Engineers: New York, NY, USA, 2015.
11. Zhang, D.; Wang, Y.; Xie, Y. Investigation into Off-Design Performance of a S-CO2 Turbine Based on Concentrated Solar Power. *Energies* **2018**, *11*, 3014. [CrossRef]
12. Baltadjiev, N.D.; Lettieri, C.; Spakovszky, Z.S. An Investigation of Real Gas Effects in Supercritical CO_2 Centrifugal Compressors. *J. Turbomach. Trans. ASME* **2015**, *137*. [CrossRef]
13. Conboy, T.; Wright, S.; Pasch, J.; Fleming, D.; Rochau, G.; Fuller, R. Performance Characteristics of an operating Supercritical CO_2 Brayton Cycle. *J. Eng. Gas Turbines Power* **2012**, *134*, 941–952. [CrossRef]
14. Lettieri, C.; Baltadjiev, N.; Casey, M.; Spakovszky, Z. Low-Flow-Coefficient Centrifugal Compressor Design for Supercritical CO_2. *J. Turbomach. Trans. ASME* **2014**, *136*. [CrossRef]
15. Lopez, A.; Monje, B.; Sanchez, D.; Chacartegul, R.; Sanchez, T. *Effect of Turbulence and Flow Distortion on the Performance of Conical Diffusers Operating on Supercritical Carbon Dioxide*; America Society Mechanical Engineers: New York, NY, USA, 2013.
16. Noughabi, A.K.; Sammak, S. Detailed Design and Aerodynamic Performance Analysis of a Radial-Inflow Turbine. *Appl. Sci.* **2018**, *8*, 2207. [CrossRef]
17. Deng, Q.; Shao, S.; Fu, L.; Luan, H.; Feng, Z. An Integrated Design and Optimization Approach for Radial Inflow Turbines-Part II: Multidisciplinary Optimization Design. *Appl. Sci.* **2018**, *8*, 2030. [CrossRef]
18. Zweifel, O. The Spacing of Turbo-Machine Blading, Especially with Large Angular Deflection. *Brown Boveri Rev.* **1945**, *32*, 436–444.
19. Horlock, J.H. *Axial Flow Turbines*; Butterworth-Heinemann: Oxford, UK, 1966.
20. Chen, T.Y.; Liao, Y.T.; Cheng, C.C. Development of small wind turbines for moving vehicles: Effects of flanged diffusers on rotor performance. *Exp. Therm. Fluid Sci.* **2012**, *42*, 136–142. [CrossRef]
21. Mohamed, M.H. Impacts of solidity and hybrid system in small wind turbines performance. *Energy* **2013**, *57*, 495–504. [CrossRef]
22. Eboibi, O.; Danao, L.A.M.; Howell, R.J. Experimental investigation of the influence of solidity on the performance and flow field aerodynamics of vertical axis wind turbines at low Reynolds numbers. *Renew. Energy* **2016**, *92*, 474–483. [CrossRef]
23. Gao, K.; Xie, Y.; Zhang, D. Effects of rotor solidity and leakage flow on the unsteady flow in axial turbine. *Appl. Therm. Eng.* **2018**, *128*, 926–939. [CrossRef]
24. Simpson, A.T.; Spence, S.W.T.; Watterson, J.K. Numerical and Experimental Study of the Performance Effects of Varying Vaneless Space and Vane Solidity in Radial Turbine Stators. *J. Turbomach. Trans. ASME* **2013**, *135*. [CrossRef]

25. Pereiras, B.; Takao, M.; Garcia, F.; Castro, F. *Influence of the Guide Vanes Solidity on the Performance of a Radial Impulse Turbine with Pitch-Controlled Guide Vanes*; America Society Mechanical Engineers: New York, NY, USA, 2011.
26. Dong, B.S.; Xu, G.Q.; Li, T.T.; Quan, Y.K.; Zhai, L.J.; Wen, J. Numerical prediction of velocity coefficient for a radial-inflow turbine stator using R123 as working fluid. *Appl. Therm. Eng.* **2018**, *130*, 1256–1265. [CrossRef]
27. Ji, G.H. *Turbo-Expander*, 1st ed.; China Machine Press: Beijing, China, 1982.
28. Brennen, C.E. *Hydrodynamics of Pumps*; Cambridge University Press: Cambridge, UK, 2011.
29. Deng, Q.; Shao, S.; Fu, L.; Luan, H.; Feng, Z. An Integrated Design and Optimization Approach for Radial Inflow Turbines-Part I: Automated Preliminary Design. *Appl. Sci.* **2018**, *8*, 2038. [CrossRef]
30. Menter, F.R. 2-Equation Eddy-Viscosity Turbulence Models for Engineering Applications. *AIAA J.* **1994**, *32*, 1598–1605. [CrossRef]
31. Roache, P.J. Perspective—A Method for Uniform Reporting of Grid Refinement Studies. *J. Fluids Eng. -Trans. ASME* **1994**, *116*, 405–413. [CrossRef]
32. Paudel, S.; Saenger, N. Grid refinement study for three dimensional CFD model involving incompressible free surface flow and rotating object. *Comput. Fluids* **2017**, *143*, 134–140. [CrossRef]
33. Kim, S.G.; Lee, J.; Ahn, Y.; Lee, J.I.; Addad, Y.; Ko, B. CFD investigation of a centrifugal compressor derived from pump technology for supercritical carbon dioxide as a working fluid. *J. Supercrit. Fluids* **2014**, *86*, 160–171. [CrossRef]
34. Odabaee, M.; Sauret, E.; Hooman, K. CFD Simulation of a Supercritical Carbon Dioxide Radial-Inflow Turbine, Comparing the Results of Using Real Gas Equation of Estate and Real Gas Property File. *Appl. Mech. Mater.* **2016**, *846*, 85–90. [CrossRef]
35. Ameli, A.; Uusitalo, A.; Turunen-Saaresti, T.; Backman, J. Numerical Sensitivity Analysis for Supercritical CO_2 Radial Turbine Performance and Flow Field. In Proceedings of the 4th International Seminar on Orc Power Systems, Milan, Italy, 13–15 September 2017; Volume 129, pp. 1117–1124.

 *applied sciences*

Article

Aerodynamic Optimization Design of a 150 kW High Performance Supercritical Carbon Dioxide Centrifugal Compressor without a High Speed Requirement

Dongbo Shi and Yonghui Xie *

State Key Laboratory for Strength and Vibration of Mechanical Structures, School of Energy and Power Engineering, Xi'an Jiaotong University, Xi'an 710049, China; sdb_xjtu@126.com
* Correspondence: yhxie@mail.xjtu.edu.cn; Tel.: +86-029-82664443

Received: 22 February 2020; Accepted: 16 March 2020; Published: 19 March 2020

Featured Application: In this research, a design-optimization method of a supercritical carbon dioxide compressor with high performance without a high speed requirement is proposed, which can improve the economy and reliability of the application system.

Abstract: Supercritical carbon dioxide (S-CO$_2$) Brayton cycle technology has the advantages of excellent energy density and heat transfer. The compressor is the most critical and complex component of the cycle. Especially, in order to make the system more reliable and economical, the design method of a high efficiency compressor without a high speed requirement is particularly important. In this paper, thermodynamic design software of a S-CO$_2$ centrifugal compressor is developed. It is used to design the 150 kW grade S-CO$_2$ compressor at the speed of 40,000 rpm. The performance of the initial design is carried out by a 3-D aerodynamic analysis. The aerodynamic optimization includes three aspects: numerical calculation, design software and the flow part geometry parameters. The aerodynamic performance and the off-design performance of the optimal design are obtained. The results show that the total static efficiency of the compressor is 79.54%. The total pressure ratio is up to 1.9. The performance is excellent, and it can operate normally within the mass flow rate range of 5.97 kg/s to 11.05 kg/s. This research provides an intelligent and efficient design method for S-CO$_2$ centrifugal compressors with a low flow rate and low speed, but high pressure ratio.

Keywords: supercritical carbon dioxide; centrifugal compressor; aerodynamic optimization design; numerical simulation

---

1. Introduction

Energy, environment and development are three major themes we are facing today. The extensive use of fossil energy has posed a great threat to the living space of mankind. CO$_2$ is a low-cost fluid with a low critical point (31.1 °C and 7.38 MPa). It is non-toxic and non-combustible and has great thermal stability, physical properties and safety [1]. The working temperature of the S-CO$_2$ Brayton cycle is above the critical temperature of carbon dioxide. Additionally, S-CO$_2$ has good transitivity and fast mobility, and its density is close to that of liquid. So, it can make the pressure of the fluid high [2,3]. Meanwhile, the unique properties of microsupercritical CO$_2$ can reduce compressor power consumption and improve cycle efficiency. Therefore, S-CO$_2$ Brayton cycle technology has been adopted in thermal power, nuclear power, solar power generation and so on [4]. As a key component, compressor works near the critical point of carbon dioxide. The local condensation is easy to occur, so the design of compressor is very difficult.

At present, the SNL (Sandia National Laboratories), TIT (Tokyo Institute of Technology) and KIER (Korea Institute of Energy Research) have designed and completed the S-CO$_2$ compressor equipment used in the experimental system. Table 1 summarizes the key parameters of the S-CO$_2$ centrifugal compressor in the test system of the three research institutions [5–8].

Table 1. The key parameters of the S-CO$_2$ centrifugal compressor.

Parameters	SNL [5,6]	TIT [7]	KIER [8]
Power/kW	50.2	10 (Cyclic generating capacity)	65.4
Speed/r·min^{-1}	75,000	100,000	70,000
Flow rate/kg·s^{-1}	3.46	1.20	3.20
Inlet temperature/°C	32	35	36
Inlet pressure/MPa	7.69	8.23	7.91
Impeller diameter/mm	37.3	30.0	53.0
Design efficiency/%	67	60	70

It can be seen from Table 1 that, due to the high density characteristics of S-CO$_2$, the centrifugal compressors designed by various research institutions have a smaller diameter, compact structure and higher speed. At present, the design of S-CO$_2$ compressor is not perfect. The design of the diffuser, bearing, seal and other auxiliary parts is difficult. Therefore, the S-CO$_2$ compressor is usually not able to reach the design speed in the actual operation. The experimental speed is far below the design value. For instance, in the experiment of TIT, the speed of turbine compressor sets is only 55,000 r/min [7]. In the experiment of KIER, the speed of turbine compressor sets is 30,000 r/min [8]. Moreover, the internal flow field of the compressor is far from the design point. It will lead to greater efficiency and power changes. For example, when the system of TIT is running, the centrifugal compressor efficiency is only 48%, and the circulating power generation capacity is only 0.11 kW [9]. Therefore, a series of problems such as aerodynamic design, actual operation rule, bearing and gas seal structure design, diffuser matching problem and material property changes of the S-CO$_2$ centrifugal compressor need further study. In particular, it is necessary to study the design method of a high efficiency compressor at low design speed.

In recent years, the mechanism of the S-CO$_2$ centrifugal compressor has been increasingly studied by scholars of various countries. The design of the S-CO$_2$ centrifugal compressor with a low speed coefficient is studied by Lettieri et al. of MIT (Massachusetts Institute of Technology) [10]. It is found that the use of the vane diffuser can improve the efficiency of the compressor in this case. Monje et al. [11] studied the design method of the compressor in the S-CO$_2$ Brayton cycle system, and the key parameter selection of the one-dimensional design program and multidimensional design method are introduced. Budinis et al. [12] have carried on the design and the analysis to the control system of the S-CO$_2$ compressor, and studied the variable operating curves and the surge control methods in detail. Shao et al. [13] of Dalian University of Technology introduced the concept of 'condensate margin' in the design process of the S-CO$_2$ centrifugal compressor to evaluate the working fluid state of the impeller inlet.

In summary, the compressor works in the microsupercritical point. Thus, the change of S-CO$_2$ physical property depends on the inlet conditions to some extent. The change of density with the pressure is also different from the ideal gas. Therefore, the existing scientific principles and the formulas related to avoiding surge may no longer be applicable to this system. At present, the compressor aerodynamic design lacks the empirical parameter and the range of estimated parameter corresponding to S-CO$_2$. This makes the design more difficult. In addition, the design cycle is greatly increased, and designers often need a lot of experience. Especially, high speed can improve the efficiency of compressor. However, it will greatly improve the design difficulty and cost of the high-speed motor. It will also affect the reliability of bearing and system. Based on the research status and difficulties mentioned above, the thermal design of a 150 kW high performance S-CO$_2$ centrifugal compressor without a high speed requirement is carried out using the design software. The aerodynamic analysis, aerodynamic optimization and off-design performance analysis are also carried out.

2. Design of the S-CO$_2$ Compressor

2.1. Design Method

Based on our previous design experience and methods [14,15], the thermodynamic design software of S-CO$_2$ centrifugal compressor (S-CO$_2$CPTD) was developed. The software was adopted in the compressor thermodynamic design of this research. The software has five main functions: data input, core calculation, data output, drawing and secondary exploration. The software is designed by Visual Basic 6.0 and Intel Fortran 2017.

Based on 1-D flow theory, the following main equations were used in the core calculation.

(1)　Euler Equation

According to the energy transformation and conservation law, the mechanical energy transferred to rotor blades is converted into fluid energy. Thus, the energy gained by one kilogram of fluid is:

$$h_{th} = \frac{1}{g}(c_{2u}u_2 - c_{1u}u_1) \tag{1}$$

where h_{th} (J/kg) is the energy head gained by one kilogram of fluid in the impeller blade channel, g (m/s^2) is the acceleration of gravity, c (m/s) is the absolute speed and u (m/s) is the circumferential speed. The subscript 1 represents the impeller inlet, the subscript 2 represents the impeller outlet and the subscript u represents the circumferential component.

(2)　Energy Equation

The total power of per kilogram of fluid consumed in centrifugal compressor stages is considered to consist of three parts: the work of the impeller on the fluid, the loss of internal leakage and the loss of impeller resistance. Based on the energy equation, the work of the impeller on fluid is converted to the fluid energy.

$$h_{tot} = h_{th} + h_{df} + h_t \tag{2}$$

where h_{tot} (J/kg) is the total energy head, h_{df} (J/kg) is the loss of impeller resistance and h_t (J/kg) is the loss of internal leakage.

In Formula (3), the theoretical energy head is converted to the working substance in the form of mechanical energy. The loss of internal leakage and the loss of impeller resistance are converted to fluid in the form of heat. Therefore, the total energy equation can be written as:

$$h_{tot} = \frac{c_p}{A}(T_2 - T_1) + \frac{c_2^2 + c_1^2}{2g} \tag{3}$$

where c_p (J/(kg·K)) is the specific heat capacity at constant pressure, A is the thermal equivalent of work (J/cal) and T (K) is the temperature.

(3)　Bernoulli Equation

For the whole compressor stage, the work of the impeller on the fluid is converted to the following three parts: (1) Improving the static pressure energy of fluid. (2) Improving the kinetic energy of fluid, but in general, the kinetic energy increases little, and is often negligible. (3) Overcoming the flow loss of working fluids in stages.

By adopting the Bernoulli equation in the impeller, the following equation can be obtained:

$$h_{th} = \int_1^2 \frac{dp}{\gamma} + \frac{c_2^2 - c_1^2}{2g} + h_{f1} \tag{4}$$

where p (Pa) is the pressure, γ (N/m^3) is the specific weight and h_{f1} (J/kg) is the flow loss of fluid flowing in the impeller.

When adopting the Bernoulli equation in the diffuser, h_{th} is zero:

$$\frac{c_2^2 - c_3^2}{2g} = \int_2^3 \frac{dp}{\gamma} + h_{f2} \tag{5}$$

where h_{f2} (J/kg) represents the flow loss of fluid flowing in the diffuser. The subscript 2 represents the diffuser inlet and 3 represents the diffuser outlet.

(4)   Key geometric design Equations

The thermal design is mainly to calculate the flow in the centrifugal compressor impeller to determine the geometric parameters. According to the design conditions and estimated parameters, the impeller outer diameter can be obtained:

$$D = \sqrt{\frac{\dot{m}}{\rho_2 u_{2r} \pi \tau_1 \tau_2}} \tag{6}$$

where $\dot{m}$ (kg/s) is the mass flow rate, ρ (kg/m^3) is density, τ_1 is the blocking factor of the impeller outlet, $\tau_2 = l_2/D$ is the relative width of the impeller outlet and l_2 (mm) is the outlet blade height of impeller. The subscript r represents the radial component.

The inlet blade height of the impeller can be obtained from the following equation:

$$l_1 = \frac{(1 + \tau_4) \cdot q_1}{\pi \tau_3 D' u_{1r}} \tag{7}$$

where τ_4 is the leakage loss coefficient of the impeller cover, q (m^3/s) is the volume flow rate, τ_3 is the blocking factor of the impeller inlet and D' (mm) is the outer diameter of impeller inlet section.

In this study, the radial linear design method [16] was adopted for the impeller blade, and the cylindrical parabola geometric design method was adopted for the blade profile, as shown in Figure 1. In Figure 1, m is the thickness of the blade outlet, n is the thickness of the blade inlet and z' is the axial length of blade at different radius positions. The detailed methods and specific equations are given by Reference 16. The diffuser is in the form of the airfoil blade, and the optimal design is realized by optimizing the profile and the geometry angle of the inlet and outlet.

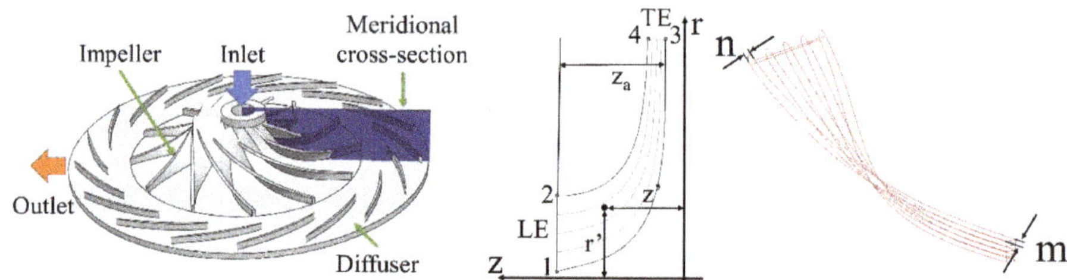

Figure 1. Geometric design diagram of the impeller blade.

2.2. The Intial Design

The key parameters of the initial design of compressor are shown in Table 2. The design speed was 40,000 rpm. There were 12 main blades and 12 splitter blades. The rotor's meridian face profile and 3-D model are shown in Figure 2. On the basis of the initial design, the Case A (9 main blades + 9 splitter blades) and the Case B (12 main blades) were designed to analyze the influence of blade form and number on compressor performance. The geometric parameters and blade profiles of the initial design, Case A and Case B were the same.

Table 2. Key parameters of the initial design of the compressor.

Key Parameters		Value	Unit
Thermodynamic parameters	Speed	40,000	rpm
	Inlet pressure	7.8	MPa
	Inlet temperature	309.15	K
	Mass flow rate	6.66	kg/s
Geometric parameters	Wheel diameter	76	mm
	Inlet diameter	15	mm
	Blade height (inlet)	11.05	mm
	Blade height (outlet)	2.4	mm
	Tip clearance	0.24	mm
	Inlet geometric angle	50	°
	Outlet geometric angle	30	°
	Leading edge sweep angle	0	°
	Axial length	15	mm
	Blade thickness	1.0	mm
	Main blades	12	pc.
	Splitter blades	12	pc.

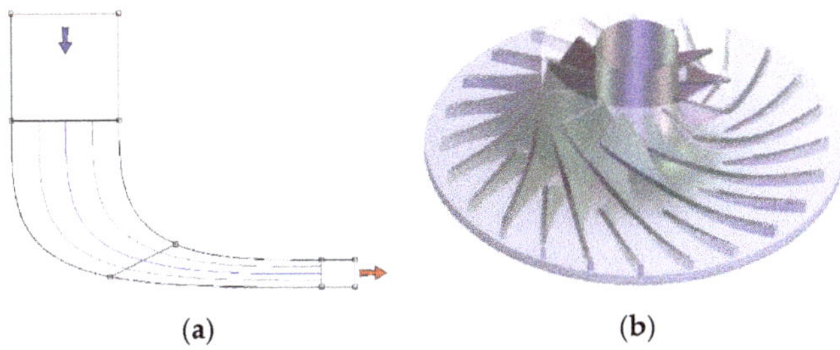

Figure 2. The initial design of the S-CO$_2$ centrifugal compressor: (**a**) rotor's meridian face profile and (**b**) 3-D model.

2.3. Optimization Method

At present, the empirical coefficients and estimated parameters matching the thermal design of the S-CO$_2$ centrifugal compressor have not been disclosed. The size of the flow passage obtained by the 1-D flow method does not necessarily conform to the actual flow requirements. At the same time, the design and optimization period of the traditional design method is long. Therefore, based on the 1-D flow theory, a fast thermal design method of S-CO$_2$ compressor was established in this study. Combined with the high-precision three-dimensional aerodynamic analysis method, a design-optimization method based on Gauss process regression was proposed, as shown in Figure 3.

According to the thermal design results and aerodynamic design results, and based on the Gauss process regression assumption, the true isentropic efficiency of an unknown design condition y can be estimated:

$$\hat{f}_r(\mathbf{y}) = f_a(\mathbf{y}) + \Delta f_i(\mathbf{y}) \tag{8}$$

where $\Delta f_i(\mathbf{y})$ is the deviation between the thermal design $f_a(\mathbf{y})$ and CFD accurate calculation results. The calculation deviation satisfies Gauss distribution at each point in the whole design space:

$$\Delta f(\mathbf{y}) \sim N(\mu(\mathbf{y}), \sigma^2(\mathbf{y})) \tag{9}$$

The mean $\mu(\mathbf{y})$ and variance $\sigma(\mathbf{y})$ of Gaussian distribution are:

$$\begin{cases} \mu(\mathbf{y}) = \mu + \mathbf{R}_{\mathbf{y}D}\mathbf{R}_{DD}^{-1}(\Delta f_D - 1\mu) \\ \sigma(\mathbf{y}) = \sigma^2(1 - \mathbf{R}_{\mathbf{y}D}\mathbf{R}^{-1}{}_{DD}\mathbf{R}_{D\mathbf{y}}) \end{cases} \tag{10}$$

where D is the training set, $\mathbf{R}_{\mathbf{y}D}$ and $\mathbf{R}_{D\mathbf{y}}$ are the covariance vector of the calculated condition and the new sampling point, $\mathbf{R}_{DD}$ is the covariance matrix and Δf_D is the deviation between the thermal design results and CFD accurate calculation results of all samples in the training set.

Therefore, the real isentropic efficiency (aerodynamic analysis result) of an unknown design condition y is estimated as follows:

$$\begin{cases} \breve{f}_r(\mathbf{y}) = f_a(\mathbf{y}) + \mu(\mathbf{y}) + \xi\sigma(\mathbf{y}) \\ \widehat{f}_r(\mathbf{y}) = f_a(\mathbf{y}) + \mu(\mathbf{y}) - \xi\sigma(\mathbf{y}) \end{cases} \tag{11}$$

where $\breve{f}_r(\mathbf{y})$ is the lower limit estimation, $\widehat{f}_r(\mathbf{y})$ is the upper limit estimation and ξ is the confidence constant.

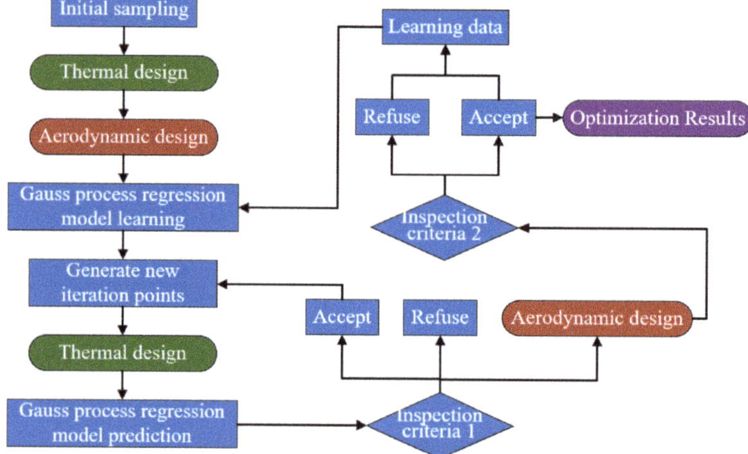

Figure 3. The design-optimization method based on the Gauss process regression.

This method has the characteristics of having fast speed and being self-adaptive. The detailed optimization theory and method can refer to Reference [17]. Key geometric parameters were used as optimization variables, including: the relative width of the impeller outlet, inlet and outlet geometric angle, leading edge sweep angle, etc. Generally, the better design of a given blade profile can be obtained by only 30 steps of iteration (i.e., the number of the aerodynamic analysis). The optimization time is only 1–2 days. According to the better design results, three aspects are optimized: numerical calculation, compressor design software and flow passage geometry parameters. The optimization process is shown in Figure 4. The numerical calculation is improved by changing the numerical calculation method and optimizing the physical property database. By changing the control parameters of numerical calculation, the stability and speed of convergence can be improved. The interpolation density of physical database is optimized to improve the calculation speed on the premise of ensuring the calculation accuracy. The compressor design software is modified by adjusting the experience coefficient, the loss model and the design method through the thermal design and Gauss process regression method. Meanwhile, the flow passage geometric parameters are optimized, such as the blade profile, geometric angle and wrap angle, the axial length of the impeller, etc. In addition, the matching of diffuser and impeller should be considered. Iterating the optimal process until the optimal design is achieved.

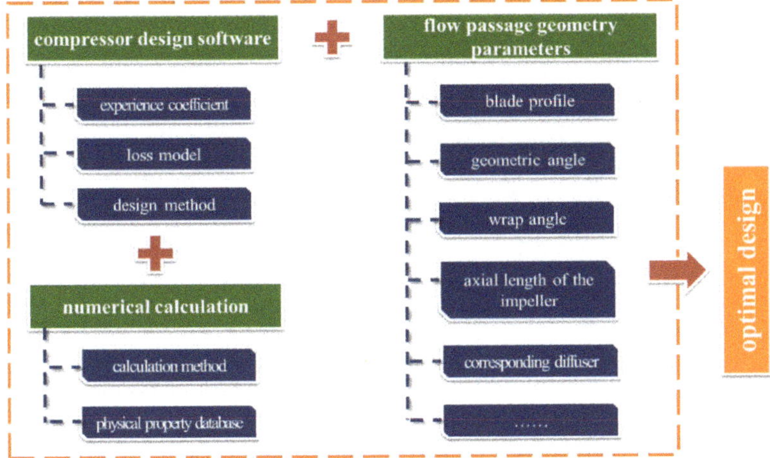

Figure 4. The optimization process.

2.4. The Optimal Design

The key parameters of the optimal design of the compressor are shown in Table 3. The design speed is 40,000 rpm. There were 12 main blades and 17 diffuser blades. Figures 5 and 6 respectively show the profile and 3-D model of the impeller and diffuser in the optimal design.

Table 3. Key parameters of the optimal design of the compressor.

Key Parameters		Value	Unit
Thermodynamic parameters	Speed	40,000	rpm
	Inlet pressure	7.8	MPa
	Inlet temperature	309.15	K
	Mass flow rate	6.66	kg/s
Geometric parameters of impeller	Wheel diameter	80	mm
	Inlet diameter	15	mm
	Blade height (inlet)	10.5	mm
	Blade height (outlet)	2.25	mm
	Tip clearance	0.25	mm
	Inlet geometric angle	50	°
	Outlet geometric angle	30	°
	Leading edge sweep angle	0	°
	Axial length	15	mm
	Blade thickness	1.0	mm
	Blades number	12	pc.
Geometric parameters of diffuser	Internal diameter	104	mm
	External diameter	126	mm
	Inlet geometric angle	20	°
	Outlet geometric angle	30	°
	Blade height	2.25	mm
	Number of blades	17	pc.

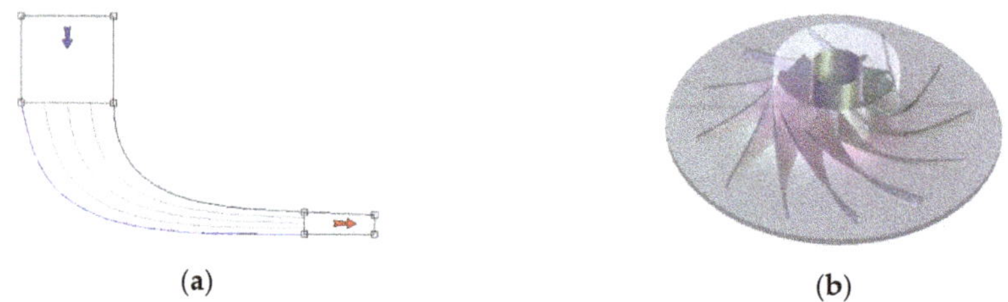

Figure 5. The impeller of the optimal design of the S-CO$_2$ centrifugal compressor: (**a**) rotor's meridian face profile and (**b**) 3-D model of the impeller.

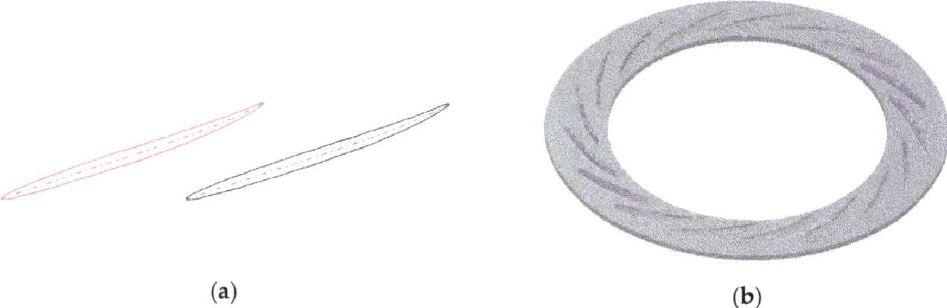

Figure 6. The diffuser of the optimal design of the S-CO$_2$ centrifugal compressor: (**a**) blade profile on the B2B cross section and (**b**) 3-D model of the diffuser.

3. Numerical Methods

3.1. Boundary Conditons

In this research, NUMECA—FineTurbo was used to solve the three-dimensional N–S equations. The single impeller–diffuser flow passage was used as the calculation model. S-CO$_2$ was used as the working substance. According to the literature [18] and related data (NIST database), the thermal physical data of microsupercritical and microsubcritical CO$_2$ were integrated. The thermal physical property continuous function was constructed by the Kriging surrogate model. The points near the critical point were locally encrypted because of the sharp change of physical properties near the critical point. The dynamic database of thermal physical properties of S-CO$_2$ was established for invocation, which can ensure the accuracy and convergence of calculation. The pressure range of working medium was set to 1–20 MPa, and the temperature range was set to 273.15–600 K. The fluid model included the parameters of all the phases of CO$_2$ in the above pressure and temperature range. The Shear Stress Transport (extended wall function) turbulence model was selected. Some other scholars [19–21] have also used the turbulence model in the analysis of centrifugal compressor, and obtained reasonable results. For the impeller inlet, the total temperature was 309.15 K and the total pressure was 7.8 MPa. k (5 m^2/s^2) and epsilon (30,000 m^2/s^3) were selected as turbulent quantities. For the diffuser outlet, the flow rate was set as 6.66 kg/s, and the influence of backflow was taken into consideration. The impeller fluid domain had a rotational speed of 40,000 rpm around the Z axis. The cross section between the impeller outlet and diffuser inlet was set as a coupling interface with conservative coupling by the pitchwise row mixed model. The diffuser wall was set as absolutely static, and the impeller wall was set as relatively static. The shroud and hub were set as adiabatic, and the conditions of non-slip flow were satisfied. The residual less than 1×10^{-6} and the iterative deviation less than 0.1% of the outlet temperature and impeller blade torque were considered as the convergence conditions.

3.2. Mesh and Grid Independence

Figure 7 shows the grid schematic diagram of the impeller–diffuser fluid domain. In this paper, NUMECA—AutoGrid5 was used to mesh. The H type mesh was used in the inlet and outlet extension sections. For higher mesh quality, the O type mesh was used for blade meshing. The grids of the tip clearance and near the wall were refined to obtain accurate flow parameters. The value of Y^+ near the wall was about 1, which met the calculation requirements of the SST turbulence model.

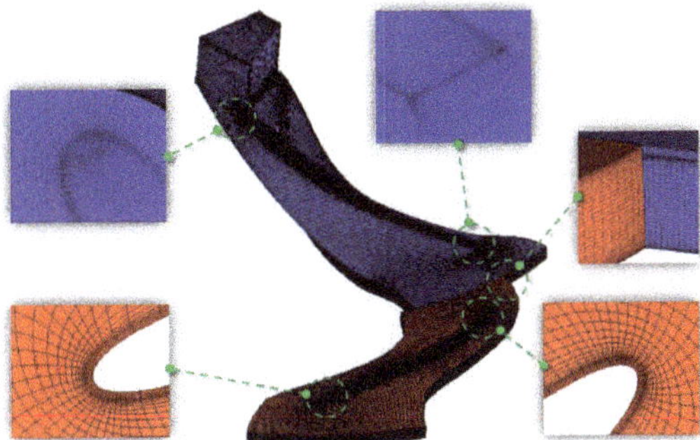

Figure 7. The grid schematic diagram of the impeller–diffuser fluid domain.

In order to effectively utilize computing resources, the grid independence verification was carried out. The optimal design of the compressor was taken as an example. The grid independence is shown in Figure 8. With the increase of the grid number, the accuracy of the calculation model also increased. More details of flow loss could be captured in the numerical calculation, so the efficiency was gradually reduced. When the grid number increased from 800,000 to 1,600,000, the relative change of total static efficiency was less than 0.2%. So, when the grid number was more than 800,000, the calculation model could meet the demand of the calculation accuracy.

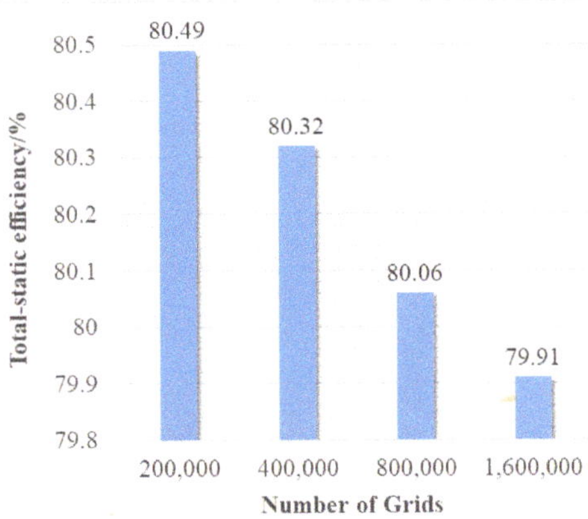

Figure 8. The grid independence.

The total static efficiency was determined by the output power, mass flow rate, isentropic enthalpy rise and inlet velocity. The isentropic enthalpy drop and the total static efficiency are shown in the Formulas (12) and (13), respectively.

$$\Delta h_s = c_p T_{inlet}\left[1 - \left(\frac{p_{out}}{p_{inlet}}\right)^{\frac{\kappa-1}{\kappa}}\right] \tag{12}$$

$$\eta_{ts} = \frac{P}{\dot{m}(\Delta h_s + c_{inlet}^2/2)} \tag{13}$$

where κ is the adiabatic exponent and P (W) is the power. The subscript inlet represents the compressor inlet and the subscript outlet represents the compressor outlet.

3.3. Numerical Validation

In order to verify the accuracy of the numerical method, the single-stage compressor model of SNL was adopted. The key geometric parameters of the model are summarized in Table 4. The case of the total pressure of 7.69 MPa, total temperature of 305.95 K and rotation speed of 50,000 rpm was analyzed. The verification results were compared with the numerical results from Rinaldi et al. [22] and the experimental results from Wright et al. [5], as shown in Figure 9. The calculated efficiency curve was quite close to that from Rinaldi et al., and shows the same trend. However, there was a certain error between the setting of boundary conditions in the numerical calculation and the actual conditions in the experiment. Therefore, the difference between the numerical calculation and experiment was also acceptable. In general, the numerical method in this study was accurate.

Table 4. The geometric parameters of the SNL S-CO$_2$ centrifugal compressor [5].

Parameters	Value
Inlet hub radius/mm	2.5375
Inlet blade height/mm	6.8345
Outlet blade height/mm	1.7120
Impeller diameter/mm	18.6817
Blade thickness/mm	0.762
Number of main blades/pc.	6
Number of splitter blades/pc.	6
Number of diffuser blades/pc.	17
Tip clearance/mm	0.254
Radial gap between impeller and diffuser/mm	0.3
Inlet blade angle at tip/°	50
Outlet blade back sweep angle/°	−50
Outlet vaned diffuser angle/°	71.5
Diffuser blade divergence angle/°	13.17

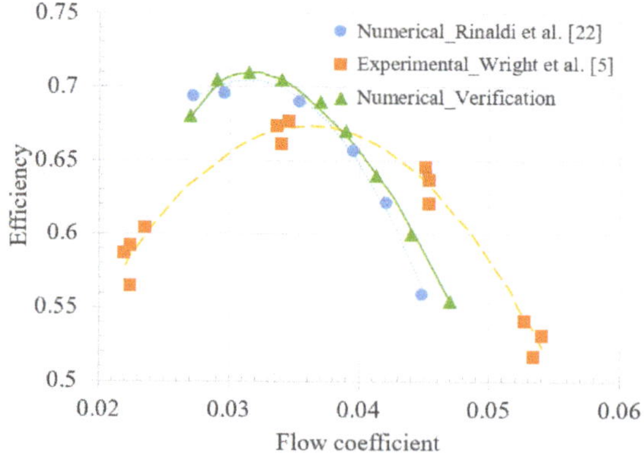

Figure 9. The numerical validation.

4. Results and Discussion

4.1. The Initial Design

Figure 10 shows the velocity vector and partial enlargement of the 50% blade height section of the compressor initial design.

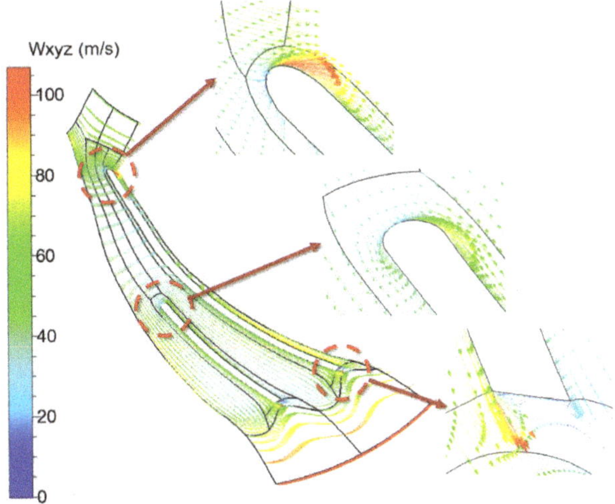

Figure 10. The velocity vector distribution of the 50% blade height section of the initial design.

It can be seen from the diagram that there was no obvious flow separation phenomenon inside the impeller. However, the flow inside the impeller was not uniform. When the fluid flowed along the main blade to 50% of the chord length (the leading edge (LE) of the splitter blade), the flow velocity of the suction side (SS) increased rapidly, while the fluid on the pressure side (PS) still flowed at a lower velocity. This led to a larger pressure gradient between the PS and SS. Thus, the working fluid on the PS would leak through the tip clearance to the SS. It would lead to a significant reduction in compressor performance inevitably.

Figure 11 shows the pressure distribution of the impeller blades surface in detail. Generally speaking, the PS pressure was greater than SS pressure. The pressure distribution on the splitter blade surface was basically consistent with that on the corresponding area of the main blade. Under the action of centrifugal force, the pressure of CO_2 increased gradually along the blade profile. There was a larger pressure gradient at the trailing edge (TE). There was no reverse pressure region in the pressure distribution. That means there was no flow separation phenomenon on the blade surface. There was a small low-pressure region at the LE of the impeller blade. It was caused by the local acceleration of the flow impact.

When the pressure in the low-pressure region of the LE is lower than the critical point, the transcritical phenomenon will occur. To some extent, it will affect the performance of the compressor. In order to display the cross critical region of the blade surface more clearly, the maximum pressure of the contour was set to 7.38 MPa (CO_2 critical pressure), as shown in Figure 12. The phenomenon was mainly caused by two factors. Firstly, the working fluid entered the compressor along the axial direction, and there was a flow acceleration phenomenon at the rotor LE. The curvature change of the molded lines at the rotor LE was the largest. Thus, the gradient of the velocity increase was correspondingly the largest. At the corner position of the LE on the top of the blade, the accelerating fluid on both sides of the blade would cause an 'ejection' effect. The gradient of the velocity increase was the largest at this position, so the fluid velocity increased sharply. It would produce an obviously low pressure and low temperature region. CO_2 entered the two-phase region and was far below the critical state. So, it would result with the possibility of 'condensation'. Secondly, the phenomenon of an 'off vortex at the LE on the top of the blade' near this position would cause an obviously low-pressure area, and the working fluid would enter the two-phase region. This is consistent with the conclusions of other scholars [23,24].

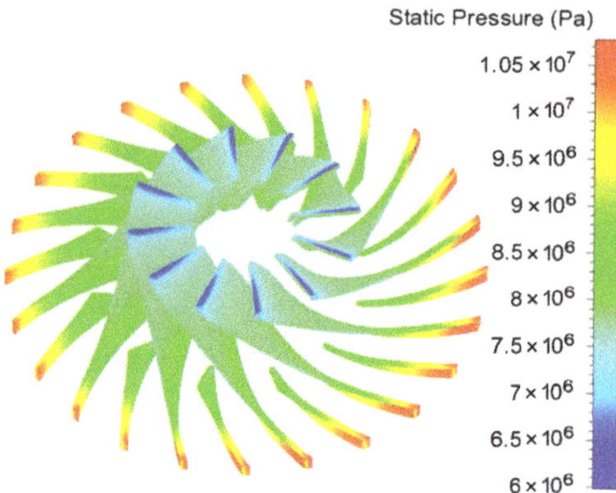

Figure 11. The pressure distribution on the surface of impeller blades.

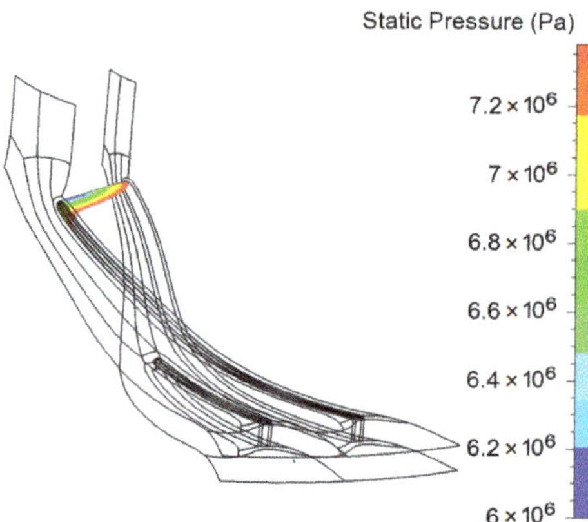

Figure 12. The cross critical region of the blade surface.

On the basis of the initial design of the compressor, this paper also contrasted and analyzed Case A and Case B. The calculated performance parameters are shown in Table 5. As can be seen from the table, compared with the initial design, although the total pressure ratio of Case B decreased from 1.67 to 1.65, the total static efficiency increased from 50.3% to 64.56% greatly. The pressure ratio of Case A was 1.67 and the total static efficiency was 63.36%. It was slightly lower than the efficiency of Case B. If the blade number is adjusted, the performance of the compressor may be equivalent to that of Case B, and may even be slightly higher. Compared with the numerical results, it was found that the addition of the splitter blade had little influence on the streamline, pressure and temperature distribution in the whole fluid domain. This is because the S-CO$_2$ compressor had a higher speed and a more compact structure than the centrifugal compressor with a conventional working medium. Therefore, it is only necessary to set the appropriate number of main blades to restrict the flow at the impeller outlet and make the outlet flow angle meet the requirements of the design value. Additionally, the compressor design with the splitter blade structure had a smaller flow capacity and larger torque. This would lead to a significant decrease in the compressor efficiency. Besides, the addition of splitter blade structure would greatly increase manufacturing difficulty and manufacturing cost. Therefore, considering the factors such as performance, machining and economy, the splitter blade structure was not considered in the following research. Overall, the compressor performance of initial design was quite different from the design value, thus the three-dimensional aerodynamic optimization is needed.

Table 5. Performance analysis of the initial design compressor.

Performance Parameter	The Initial Design (12 Main Blades + 12 Splitter Blades)	Case A (9 Main Blades + 9 Splitter Blades)	Case B (12 Main Blades)
Input power/kW	147.72	137.55	131.98
Total pressure ratio	1.67	1.67	1.65
Total static efficiency/%	50.30	63.36	64.56

4.2. The Optimal Design

Figure 13 shows the velocity vector distribution of the 50% blade height section for the compressor optimal design, and the partial enlargement of the LE and TE of rotor and diffuser blades. It can be seen from the diagram that there was no serious flow separation phenomenon and backflow phenomenon in the impeller and diffuser. The flow field was uniform. The local acceleration phenomenon was inevitable only in the LE of the impeller blade and diffuser blade. In addition, there was a small vortex produced by the impeller blade blunt TE.

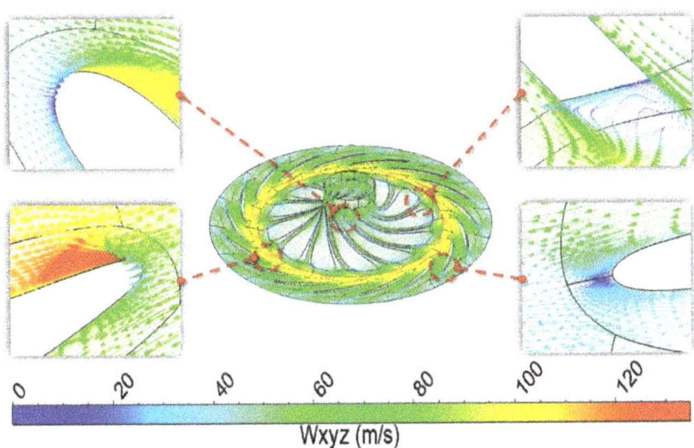

Figure 13. The velocity vector distribution of the 50% blade height section for the optimal design of the compressor.

Figure 14 shows the secondary flow velocity vector with different axial chord lengths on the S3 cross section. The vortex structure and development process of the compressor impeller can be analyzed by secondary flow velocity. For the horseshoe vortex, its intensity and influence range were obviously weaker than that of the radial inflow turbine. Only horseshoe vortex bifurcation on the PS (HVp) was found near the LE of the rotor blade, while horseshoe vortex bifurcation on the SS (HVs) was not found. Under the pressure gradient, the boundary layer was rolled up and the horseshoe vortex bifurcation was formed on the PS. As the fluid flows downstream, the influence range increased continuously. It finally merged with the passage vortex under the dissipation of the adverse pressure gradient. Additionally, the difference between the passage vortex and other vortex structures was not obvious. Figure 14a shows that there was an up passage vortex (PVu) and down passage vortex (PVd) with the same rotational direction at the tip and root of the blade on the PS, respectively. However, the intensity was small and the influence range was narrow. The up passage vortexes were generated mainly by the interaction of three parts. The first part was the scraping flow produced by the relative motion of the upper end wall due to the high-speed rotation of the impeller. The second part was the leakage flow from the PS to the SS in the tip clearance. Another part was the effect of transverse pressure gradient. As the fluid flowed along the radial direction, the scraping effect caused by the high-speed flow of the impeller became stronger, and the vortices on the PS expanded gradually. The influence range of vortex in the down passage gradually decreased. Its position was gradually squeezed into the SS and eventually dissipated and disappeared, as shown in Figure 14b. Near the TE of the rotor blade, the pressure on the SS was basically the same as that on the PS. The leakage flow and the transverse pressure gradient disappear, and the vortex system was dissipated and disappeared, as shown in Figure 14c.

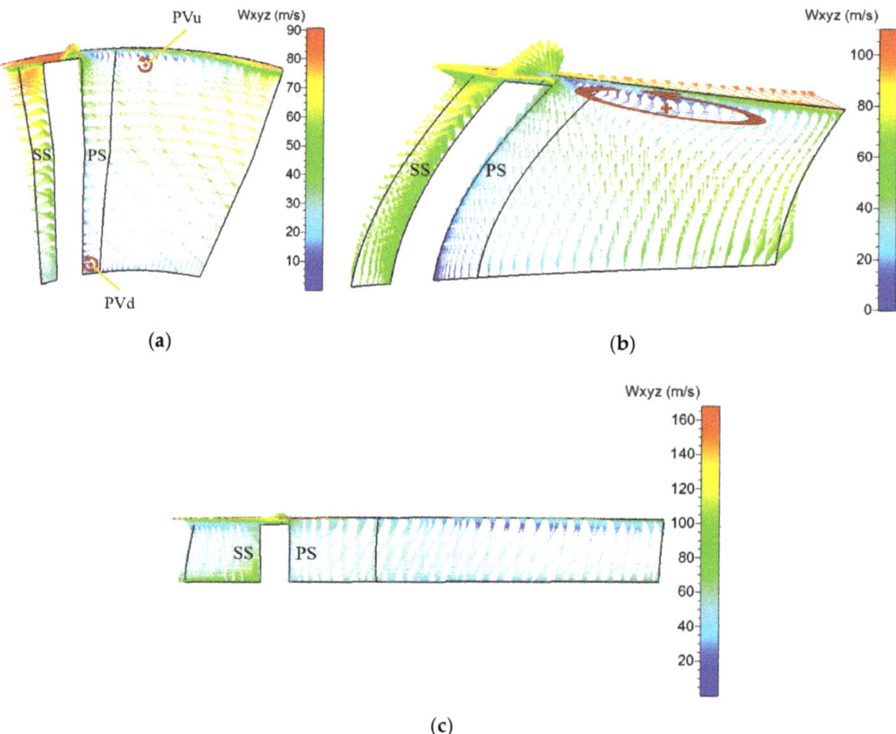

Figure 14. Second flow velocity on S3 sections of the optimal compressor impeller: (**a**) 30% axial chord length; (**b**) 40% axial chord length and (**c**) 90% axial chord length.

Figure 15 shows the pressure and temperature distribution of the 50% blade height section for the compressor optimal design. Generally speaking, the pressure and temperature of the PS were greater than that of the SS. Under the action of centrifugal force, the CO_2 pressure and temperature from the LE to the TE of the rotor gradually increased, and there was a larger gradient at the TE. The diffusing action of diffuser maximized the outlet pressure and outlet temperature. The outlet static pressure reached 14.25 MPa.

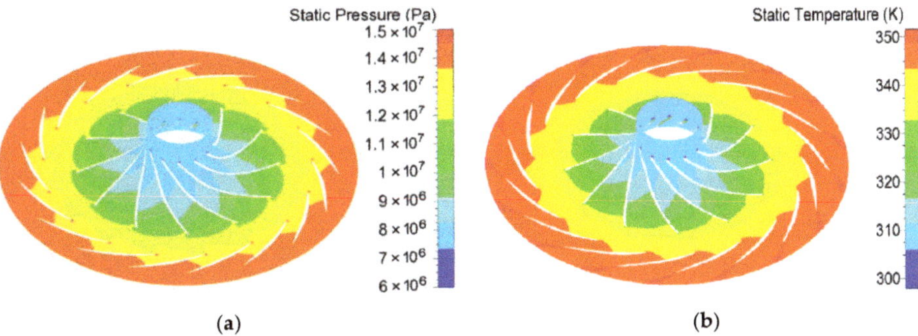

Figure 15. The pressure and temperature distribution of the 50% blade height section for the optimal design of the compressor: (**a**) pressure distribution and (**b**) temperature distribution.

Figure 16 shows the surface pressure distributions of the impeller blades and diffuser blades of the compressor optimal design. The blade surface pressure curves with different blade height sections are shown in Figure 17. In general, the pressure increased gradually from the inlet to the outlet, the pressure gradient distribution was uniform. The maximum pressure was at the diffuser outlet. Due to the decrease of the CO_2 leakage caused by the tip clearance, the surface pressure of rotor was basically the same along the blade height direction from the position of the 50% chord length to the TE. The cross critical region of the blade surface is shown in Figure 18. Similarly, the fluid in the blade tip clearance at the impeller blade LE was 'ejected'. Meanwhile, the fluid at the LE of the impeller blade would be affected by the phenomenon of an 'off vortex at the LE on the top of blade'. The simultaneous action of the two phenomena led to obviously low-pressure-temperature regions at the impeller blade LE, and the carbon dioxide entered the two-phase region. The fluid state at the corner position of the LE on the top of the blade was far below the critical state. This might cause 'condensation'. The flow in the transcritical region was very complex and the physical properties changed dramatically. This would cause great loss and affect the flow condition in the impeller. By optimizing the geometry of the flow passage, the influence range of the transcritical phenomenon in the optimal design was obviously weakened. It was mainly concentrated on the SS of the LE.

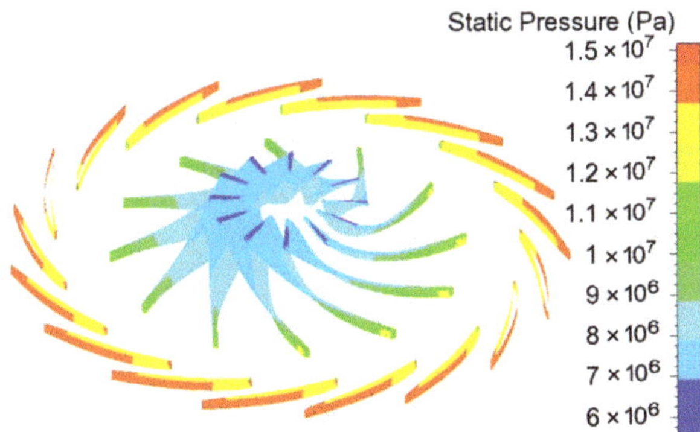

Figure 16. The pressure distribution of the blade surface.

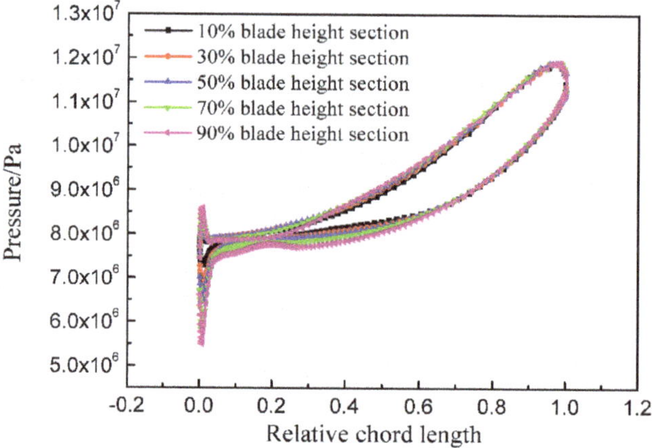

Figure 17. The pressure curves of the blade surface with different blade high cross sections.

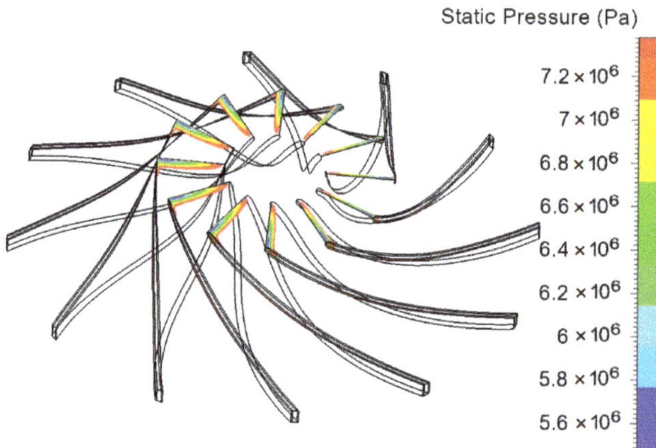

Figure 18. The cross critical region of the blade surface.

In order to study the loss of flow in the compressor more intuitively, the entropy distribution of different blade height sections, the impeller outlet and diffuser outlet are given in Figure 19. It can be found that the flow loss inside the compressor was more and more serious from the hub to the shroud. It is consistent with the gradual deterioration of the flow situation mentioned above. The entropy of the 10% blade height section had little change. The working medium was pressurized in the impeller, which is similar to the isentropic process. In the diffuser, due to the influence of the viscous boundary layer and the wake, an increase in entropy appeared on the blade surface and the extension of the outlet. At the 50% blade height section, the entropy of blade TE increased greatly under the influence of the wake. The flow loss in this region was increased. A large range of higher entropy increases occurred at the 90% blade height section near the blade top. This area is close to the wall, and the working fluid was strongly sheared by the impeller at high speed. Therefore, there was a large viscous dissipation loss. At the same time, this area was affected by the tip clearance layer. Various vortex structures were blended here, and the flow was very complicated. The entropy at the impeller outlet was evenly distributed in the circumferential direction and gradually increased along the blade height. This is because the area near the shroud was greatly affected by the tip clearance layer and the cutting of the wall surface. The entropy at the diffuser outlet was basically unchanged. There was only a slight increase in entropy on the suction side. This means that there were no backflow and secondary flow in this area. In summary, the optimal design had small flow loss and good aerodynamic performance.

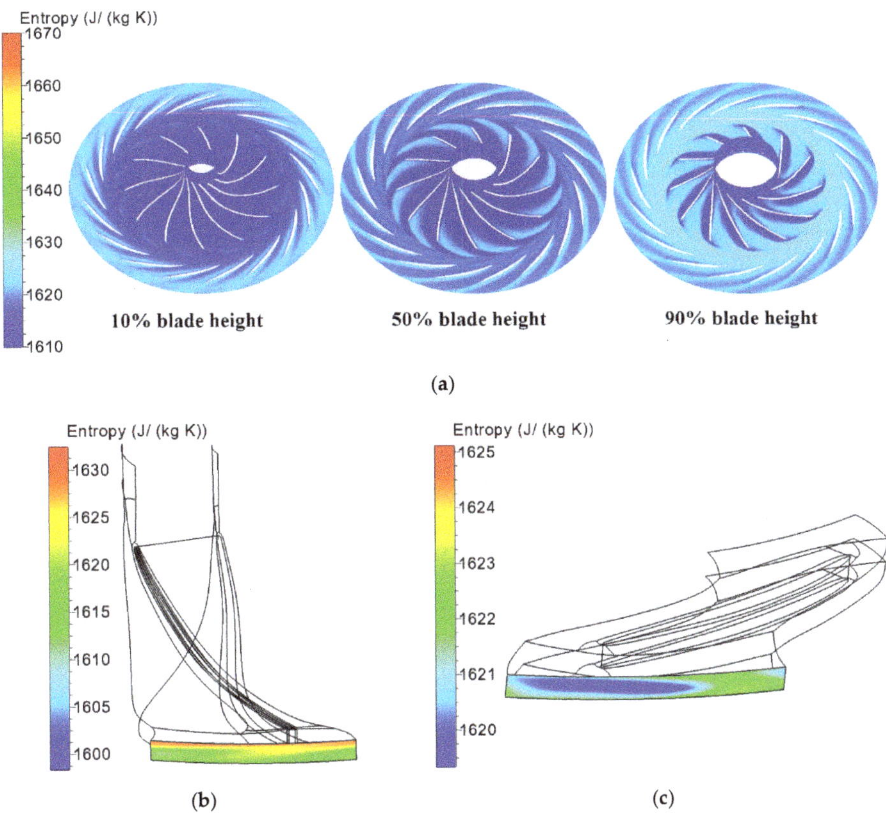

Figure 19. The entropy distribution of different sections: (**a**) different blade height sections; (**b**) impeller outlet and (**c**) diffuser outlet.

In the off-design analysis, eight kinds of outlet pressure boundary conditions were set up. The mass flow rate-total pressure ratio curve and mass flow rate-total static efficiency curve were obtained, as shown in Figures 20 and 21.

It can be seen from the figure that as the mass flow rate increased, the total pressure ratio and total static efficiency showed a decreasing trend. It is worth noting that the compressor performance was higher under the small flow condition deviating from the design condition. When the mass flow rate was 5.97 kg/s, the compressor had the highest pressure ratio and total static efficiency, 1.90% and 80% respectively. When the mass flow rate was in the range of 5.97–9.52 kg/s, the compressor performance curve changed gently. At this time, the compressor had a higher pressure ratio and efficiency. Under the condition of a mass flow rate less than 5.97 kg/s, the pressure ratio was too large, which led to the increase of outlet pressure and further caused the backflow and compressor surge. The sensitivity of mass flow rate to compressor performance was greatly increased while the mass flow rate was greater than 9.52 kg/s. The total pressure ratio and the total static efficiency decreased sharply with the increase of mass flow rate, and the blocking phenomenon was becoming more and more serious. As the mass flow rate continued to increase to 11.05 kg/s, the flow at a throat in the compressor channel reached a critical state. At this time, the flow rate was at its maximum. No matter how much the compressor back pressure was lowered, the flow rate would no longer increase. The compressor would enter the blocking condition.

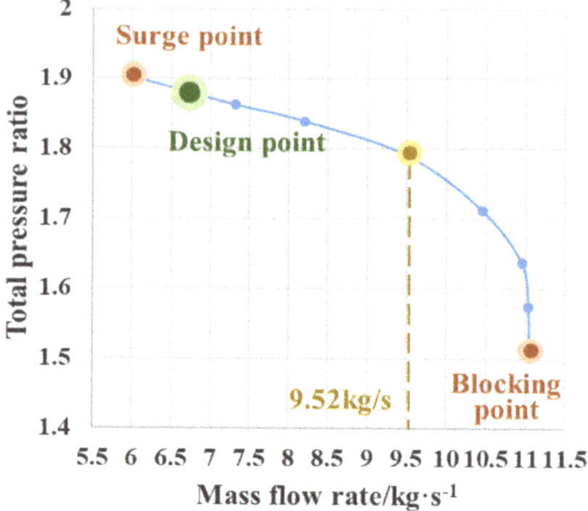

Figure 20. The compressor mass flow rate-total pressure ratio curve.

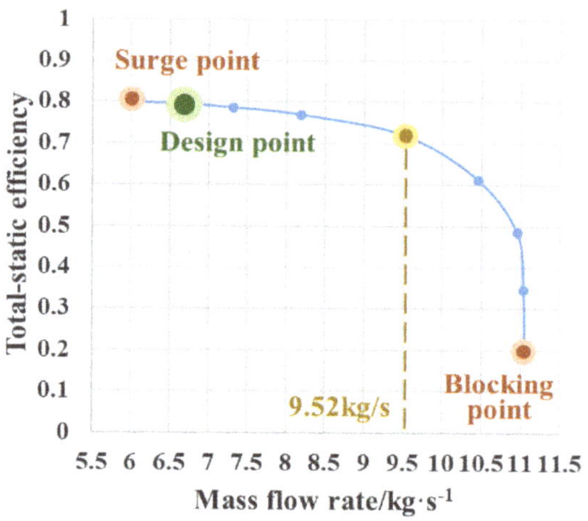

Figure 21. The compressor mass flow rate-total static efficiency curve.

The performance of the compressor optimal design in this paper was compared with the most advanced SNL centrifugal compressor in the current public literature, as shown in Table 6. Compared with the compressor of SNL, the compressor speed of the optimal design was 40,000 rpm, and the speed was almost reduced to 50%. It is worth noting that in this paper, the design speed was greatly reduced. This means that it had more testability, lower motor cost, simpler system composition and other advantages. Meanwhile, the total pressure ratio was 1.9, bigger than 1.8 of SNL. Besides, in terms of total static efficiency, it was greatly exceeded, which was 13.16% higher than the compressor of SNL. In general, the compressor designed in this study had a low speed requirement and strong comprehensive performance.

Table 6. Comparison of compressor performances.

Parameters	The Compressor Optimal Design	SNL [5,6]
Speed/rpm	40,000	75,000
Total pressure ratio	1.9	1.8
Total static efficiency	79.54	66.38

5. Conclusions

In this paper, a thermodynamic design software of the $S\text{-}CO_2$ centrifugal compressor ($S\text{-}CO_2$CPTD) was developed based on the 1-D thermal design method. The 150 kW $S\text{-}CO_2$ centrifugal compressor at the speed of 40,000 rpm was designed by using the developed software. The 3-D aerodynamic analysis and aerodynamic optimization was carried out in the compressor initial design. The concrete conclusions are as follows:

1. The initial design of compressor was done by the thermodynamic design software of the $S\text{-}CO_2$ centrifugal compressor. The power was 147.72 kW, the total pressure ratio was 1.67 and the total static efficiency was 50.30%. On the basis of the initial design of the compressor, Case A (9 main blades + 9 splitter blades) and Case B (12 main blades) were analyzed. Considering the factors such as performance, machining and economy, the splitter blade structure was not recommended for compressor design in this research.

2. According to the 3-D numerical results of variable initial designs, the software modification was made by adjusting the empirical coefficient, the loss model and the design method. The thermal design software matching the $S\text{-}CO_2$ centrifugal compressor was obtained. It made the design of the $S\text{-}CO_2$ centrifugal compressor intelligent, accurate and efficient.

3. Based on the improved numerical calculation method, the modified compressor design software and the optimized flow passage geometry parameters, the optimal design of $S\text{-}CO_2$ centrifugal compressor was obtained. The 3-D aerodynamic analysis and off-design performance analysis of the optimal design were carried out. The results show that the aerodynamic performance of the compressor was excellent at the low speed requirement of 40,000 rpm. The total pressure ratio was 1.9 and the total static efficiency was 79.54% under the design condition. The $S\text{-}CO_2$ centrifugal compressor of optimal design could run normally in the flow rate range of 5.97–11.05 kg/s. Additionally, it could run efficiently and stably in the range of 5.97–9.52 kg/s.

Author Contributions: Conceptualization, D.S. and Y.X.; investigation, D.S.; methodology, D.S.; resources, Y.X.; software, D.S. and Y.X.; supervision, Y.X.; validation Y.X.; writing—original draft preparation, D.S.; writing—review and editing, D.S. All authors have read and agreed to the published version of the manuscript.

Funding: This research received no external funding.

Conflicts of Interest: The authors declare no conflict of interest.

References

1. Dostal, V.; Hejzlar, P.; Driscoll, M.J. High-performance Supercritical Carbon Dioxide Cycle for Next-generation Nuclear Reactors. *Nucl. Sci. Tech.* **2006**, *154*, 265–282. [CrossRef]

2. Tanaka, H.; Nishiwaki, N.; Hirata, M.; Tsuge, A. Forced Convection Heat Transfer to Fluid Near Critical Point Flowing in Circular Tube. *Int. J. Heat Mass Tran.* **1971**, *14*, 739–750.

3. Brassington, D.J.; Cairn, D.N. Measurements of Forced Convective Heat Transfer to Supercritical Helium. *Int. J. Heat Mass Tran.* **1977**, *20*, 207–214. [CrossRef]

4. Zhao, X.; Lu, J.; Yuan, Y.; Dang, Y.; Gu, Y. Analysis of Supercritical Carbon Dioxide Brayton Cycle and Candidate Materials of Key Hot Components for Power Plants. *Proc. Chin. Soc. Elect. Eng.* **2016**, *36*, 154–162.

5. Wright, S.; Radel, R.; Vernon, M.; Pickard, P. *Operation and Analysis of a Supercritical CO_2 Brayton Cycle*; Sandia National Laboratories: Livermore, CA, USA, 2010.

6. Wright, S.; Pickard, P.; Radel, R. Supercritical CO_2 Brayton Cycle Power Generation Development Program and Initial Test Results. In Proceedings of the ASME Power Conference, Albuquerque, NM, USA, 21–23 July 2009.

7.   Utamura, M.; Fukuda, T.; Aritomi, M. Aerodynamic Characteristics of a Centrifugal Compressor Working in Supercritical Carbon Dioxide. In Proceedings of the 2nd International Conference on Advances in Energy Engineering, Bangkok, Thailand, 27–28 December 2011.

8.   Cho, J.; Choi, M.; Baik, Y.J.; Lee, G.; Ra, H.S.; Kim, B.; Kim, M. Development of the Turbomachinery for the Supercritical Carbon Dioxide Power Cycle. *Int. J. Energy Res.* **2016**, *40*, 587–599. [CrossRef]

9.   Utamura, M.; Hasuike, H.; Yamamoto, T. Demonstration Test Plant of Closed Cycle Gas Turbine with Supercritical CO_2 as Working Fluid. *Strojarstvo* **2010**, *52*, 459–465.

10.   Lettieri, C.; Baltadjiev, N.; Casey, M.; Spakovszky, Z. Low-Flow-Coefficient Centrifugal Compressor Design for Supercritical CO_2. *J. Turbomach.* **2014**, *136*, 081008. [CrossRef]

11.   Monje, B.; Sanchez, D.; Chacartegui, R.; Sánchez, T.; Savill, M.; Pilidis, P. Aerodynamic Analysis of Conical Diffusers Operating with Air and Supercritical Carbon Dioxide. *Int. J. Heat Fluid Flow* **2013**, *44*, 542–553. [CrossRef]

12.   Budinis, S.; Thornhill, N.F. Supercritical Fluid Recycle for Surge Control of CO_2 Centrifugal Compressors. *Comput. Chem. Eng.* **2016**, *91*, 329–342. [CrossRef]

13.   Shao, W.; Wang, X.; Yang, J.; Liu, H.; Huang, Z. Design Parameters Exploration for Supercritical CO_2 Centrifugal Compressors under Multiple Constraints. In Proceedings of the ASME Turbo Expo: Turbine Technical Conference and Exposition, Seoul, Korea, 13–17 June 2016.

14.   Wang, Y.; Shi, D.; Zhang, D.; Xie, Y. Investigation on Unsteady Flow Characteristics of a SCO_2 Centrifugal Compressor. *Appl. Sci.* **2017**, *7*, 310. [CrossRef]

15.   Shi, D.; Wang, Y.; Xie, Y.; Zhang, D. The Influence of Flow Passage Geometry on the Performances of a Supercritical CO2 Centrifugal Compressor. *Therm. Sci.* **2018**, *22*, S409–S418. [CrossRef]

16.   Feng, Z.; Deng, Q.; Li, J. Aerothermodynamic Design and Numerical Simulation of Radial Inflow Turbine Impeller for a 100kW Microturbine. In Proceedings of the ASME Turbo Expo, Reno-Tahoe, NV, USA, 6–9 June 2005.

17.   Shi, D.; Liu, T.; Xie, Y.; Zhang, D. Design and Optimization of an S-CO_2 Turbine Based on Gauss Process Regression. *Chin. J. Power Eng.* **2019**, *39*, 876–883.

18.   Shi, D.; Zhang, L.; Xie, Y.; Zhang, D. Aerodynamic Design and Off-design Performance Analysis of a Multi-Stage S-CO2 Axial Turbine Based on Solar Power Generation System. *Appl. Sci.* **2019**, *9*, 714. [CrossRef]

19.   Mojaddam, M.; Hajilouy-Benisi, A.; Moussavi-Torshizi, S.A.; Movahhedy, M.R.; Durali, M. Experimental and Numerical Investigations of Radial Flow Compressor Component Losses. *J. Mech. Sci. Technol.* **2014**, *28*, 2189–2196. [CrossRef]

20.   Nili-Ahmadabadi, M.; Hajilouy-Benisi, A.; Durali, M.; Ghadak, F. Investigation of a Centrifugal Compressor and Study of the Area Ratio and TIP Clearance Effects on Performance. *J. Therm. Sci.* **2008**, *17*, 314–323. [CrossRef]

21.   Mangani, L.; Casartelli, E.; Mauri, S. Assessment of Various Turbulence Models in a High Pressure Ratio Centrifugal Compressor with an Object Oriented CFD Code. *J. Turbomach.* **2012**, *134*, 2219–2229. [CrossRef]

22.   Rinaldi, E.; Pecnik, R.; Colonna, P. Computational Fluid Dynamics of a Radial Compressor Operating with Supercritical CO_2. *J. Eng. Gas Turbines Power* **2012**, *134*, 122301.

23.   Zhao, H.; Deng, Q.; Huang, W.; Zheng, K.; Feng, Z. Numerical Investigation on the Blade Tip Two-Phase Flow Characteristics of a Supercritical CO_2 Centrifugal Compressor. *J. Eng. Thermophys.* **2015**, *36*, 1433–1436.

24.   Rinaldi, E.; Pecnik, R.; Colonna, P. Steady State CFD Investigation of a Radial Compressor Operating with Supercritical CO_2. In Proceedings of the ASME Turbo Expo 2013: Turbine Technical Conference and Exposition, San Antonio, TX, USA, 3–7 June 2013.

Article

Design and Performance Analysis of a Supercritical Carbon Dioxide Heat Exchanger

Han Seo [1],*, Jae Eun Cha [1], Jaemin Kim [2], Injin Sah [1] and Yong-Wan Kim [1]

[1] Korea Atomic Energy Research Institute 989-111, Daedeok-daero, Yuseong-gu, Daejeon 34057, Korea; jecha@kaeri.re.kr (J.E.C.); injin@kaeri.re.kr (I.S.); ywkim@kaeri.re.kr (Y.-W.K.)

[2] National Funsion Research Institute 169-148 Gwahak-ro, Yuseong-gu Daejeon 34133, Korea; jmkim@nfri.re.kr

*   Correspondence: hanseo@kaeri.re.kr; Tel.: +82-42-868-2306

Received: 29 May 2020; Accepted: 23 June 2020; Published: 30 June 2020

Abstract: This paper presents a preliminary design and performance analysis of a supercritical CO_2 (SCO_2) heat exchanger for an SCO_2 power generation system. The purpose of designing a SCO_2 heat exchanger is to provide a high-temperature and high-pressure heat exchange core technology for advanced SCO_2 power generation systems. The target outlet temperature and pressure for the SCO_2 heat exchanger were 600 °C and 200 bar, respectively. A tubular type with a staggered tube bundle was selected as the SCO_2 heat exchanger, and liquefied petroleum gas (LPG) and air were selected as heat sources. The design of the heat exchanger was based on the material selection and available tube specification. Preliminary performance evaluation of the SCO_2 heat exchanger was conducted using an in-house code, and three-dimensional flow and thermal stress analysis were performed to verify the tube's integrity. The simulation results showed that the tubular type heat exchanger can endure high-temperature and high-pressure conditions under an SCO_2 environment.

Keywords: supercritical CO_2; heat exchanger; flow analysis; thermal stress analysis

1. Introduction

The supercritical carbon dioxide (SCO_2) Brayton cycle has been considered one of the most promising alternatives to existing power generation systems, such as the steam Rankine and gas Brayton cycles. The steam Rankine cycle can achieve high thermal efficiency due to the low pumping power, but the overall size of the system is large because the steam density of the low pressure side is lower than the atmospheric pressure. The turbine inlet temperature (TIT) of the gas Brayton cycle is higher than in the steam Rankine cycle. It can achieve a high thermal power output, but material integrity and high compression work problems still remain. The SCO_2 cycle has the advantages of both the steam Rankine and gas Brayton cycles because of its fluid characteristics. First of all, the supercritical region of the SCO_2 is readily accessible (T_c = 31.1 °C, P_c = 73.8 bar); thus, the system can be controlled easily compared to other critical state fluids. Because the SCO_2 cycle operates near the critical point, the fluid reflects both liquid and gas properties. The compression work of the SCO_2 consumes less than the conventional gas Brayton cycle. In addition, the density difference between the hot and cold sides are small. This means that the overall size of the SCO_2 power generation system can be minimized compared to the conventional power generation systems. Based on the advantages, the SCO_2 cycle can be applied to various technologies, such as concentrating power, fossil fuel, geothermal, nuclear, ship-board propulsion, waste heat recovery, etc. Therefore, the development of the SCO_2 power cycle has been extensively studied.

Studies on the SCO_2 cycle have taken place since the 1960s, but its development did not occur due to these technological advances [1,2]. In the 2000s, Dostal et al. [3] studied the SCO_2 cycle as the next power generation system for Generation IV nuclear reactors because the TIT for advanced nuclear reactors is about 550 °C, a system operating condition which is between the steam Rankine and gas

Brayton cycles. Because the SCO_2 cycle can be designed in a wide range of the TIT, the SCO_2 system has received more attention as the next power generation system. In addition, the SCO_2 cycle can also be used in various heat sources, such as fossil fuel, concentrating solar power, shipboard propulsion, waste heat recovery, and geothermal.

After validating the SCO_2 cycle as a new power generation system, numerous studies have been conducted to verify the cycle. Sandia National Laboratories designed a recompression configuration, and various experimental facilities have been developed, such as turbine and compressor performance characterizations [4–6]. The Naval Nuclear Laboratory designed and built an integrated system test (IST) for the SCO_2 Brayton cycle. The IST is a simple recuperated Brayton cycle for variable turbomachinery tests. The system demonstrated that the SCO_2 Brayton cycle was controlled throughout the entire system operation, but inherent problems related to the SCO_2 Brayton cycle were identified [7–9]. Ecogen Power Systems designed a SCO_2 power cycle (EPS-100) for exhaust heat recovery applications. The EPS-100 is the first commercial-scale SCO_2 system, and it has a 7MWe scale with multiple stages of recuperation and extraction from the heat source. The validation test of the EPS-100 was completed, and commercialization took place [10,11]. The SunShot program, which develops the SCO_2 Brayton cycle for a concentrating solar power (CSP) system, was initiated with a simple recuperated cycle [12]. The target demonstration of the SCO_2 system is 10 MWe with a 50% net thermal efficiency.

To demonstrate the major components in the SCO_2 power generation system, the SCO_2's heat exchanger and turbo-expander have been tested in a 1 MWe test loop [13,14]. A tubular-type heat exchanger was selected as the primary heater, which was connected with a commercial natural gas-fired combustor [13]. The SunShot program is the first MW-scale SCO_2 power cycle demonstration in which the TIT is higher than 700 °C. A demonstration of the SCO_2 power cycle was performed by considering the unique characteristics of the CSP system [14].

In addition, the US Department of Energy (DOE) designated SCO_2 research as a cross-cutting technology and supercritical transformational electric power (STEP) program with a collaboration between fossil, nuclear, and renewable energy [15]. The program has focused on designing, constructing, commissioning, and operating a 10-MWe SCO_2 pilot plant test facility. The detailed design of the facility and equipment is now proceeding. Fabrication and construction of a pilot test facility with a simple-cycle test will be finished at the end of 2020, and facility operation and testing with the recompression cycle is scheduled until 2022 [16,17].

In Korea, various kinds of experimental facilities have been developed and investigated for the SCO_2 power cycle [18–21]. Recently, the development and testing of the SCO_2 power cycle for a waste heat recovery system started in the Korea Atomic Energy Research Institute (KAERI). The target of the project is to develop a prototype SCO_2 power generation system for a waste heat recovery system. The project aims at developing SCO_2 core technology, such as turbomachinery and heat exchangers. A simple recuperated cycle was selected. The major components of the prototype are an SCO_2 compressor and turbine, precooler, recuperator, and waste heat recovery heat exchanger. The target of the TIT is about 430 °C. To secure the TIT, it is important to manage the high-temperature and high-pressure heat exchange technologies under a SCO_2 environment.

In this research, the SCO_2 heat exchanger that will be used in the SCO_2 power generation prototype is designed. The maximum target of the SCO_2 outlet temperature and pressure is 200 bar and 600 °C, respectively. Based on the design conditions of the SCO_2 heat exchanger, heat exchanger type selection, material selection, and tube specification based on the commercial availability were conducted. Preliminary performance analysis of the SCO_2 heat exchanger was conducted using an in-house heat exchanger code, and the flow and thermal stress analysis of the SCO_2 heat exchanger were performed using commercially available computational fluid dynamic (CFD) codes.

2. Design Considerations of Supercritical CO_2 Heat Exchanger

2.1. Operating Condition of SCO_2 Heat Exchanger

Table 1 lists the operating conditions of the SCO_2 heat exchanger. The mass flow rate of the SCO_2 is 1 kg/s, and the inlet and outlet temperatures of the SCO_2 are 300 °C and 600 °C, respectively. The outlet pressure of the SCO_2 is 200 bar. Based on the operating condition of the SCO_2 power cycle, the pressure drop of the heat exchanger was limited to 1.5 bar. The heat duty of the SCO_2 heat exchanger was 380kW. A combustion system was selected as a heat source for the SCO_2. Flue gas is composed of liquefied petroleum gas (LPG) and air. The inlet temperature of the flue gas was set at 800 °C. The inlet temperature is based on the maximum allowable stress for the heat exchanger tube. The inlet mass flow rate of the flue gas was calculated as 0.8497 kg/s.

Table 1. Operating condition of the SCO_2 heat exchanger.

Design Parameters	Operating Condition
Mass flow rate (SCO_2)	1 kg/s
Outlet pressure (SCO_2)	200 bar
Inlet & Outlet temperature (SCO_2)	300 & 600 °C
Flue gas inlet temperature	800 °C
Outlet pressure (flue gas)	atmospheric pressure

2.2. Heat Exchanger Type Selection

Heat exchanger types can be classified based on the number of working fluids, compactness, flow arrangements, and heat transfer mechanisms. Tubular, plate type, extended surface, and printed circuit heat exchanger types are typical heat exchangers used in industrial areas. Among the heat exchangers, the tubular heat exchanger is popular due to its flexibility: the core shape can be easily changed by the tube diameter, length, and arrangements. In addition, tubular heat exchangers are usually used in high-temperature and high-pressure conditions. A plate type heat exchanger consists of two flow membranes, and a number of plates are compressed or welded with a gasket. Therefore, it is not appropriate to use it in extreme operating conditions due to the possibility of leakage. Compared to the tubular and plate type heat exchangers, a higher effectiveness can be achieved by using extended surface heat exchangers. However, a high pressure drop can appear on extended surface heat exchangers. For compact size heat exchangers, printed circuit heat exchangers (PCHE) have been widely studied. The volume of a PCHE can be minimized up to 1/30 compared to conventional shell-and-tube heat exchangers with the same heat duty [22]. However, maintenance and inspection of a PCHE are difficult because these heat exchangers are manufactured by a diffusion bonding process. In addition, there are limitations in material selection for diffusion bonding processes.

The type of SCO_2 heat exchanger can be determined by the desired operating condition. Because the target operating condition of the SCO_2 heat exchanger is at a high temperature and high pressure (600 °C and 200 bar), the heat exchanger should endure high thermal stress and thermal shock. In addition, the maintenance and inspection of the heat exchanger should be easy. The tubular type heat exchanger has a low pressure drop and offers the least-risk design for the thermal shock resistance, and it has modest effectiveness. Because the flue gas was considered as the heat source, the pressure drop on the hot side should also be minimized. The flexible design of the tubular heat exchanger can offer a pressure drop on the flue gas side. Therefore, the tubular type was selected as the SCO_2 heat exchanger.

2.3. Heat Exchanger Material Selection

Heat exchanger material selection is based on a combination of cost, moderate properties under the operating condition, fabricability, and availability. In addition, candidate materials are required to have good corrosion, oxidation, carburization, and brittleness resistance under the SCO_2 condition. Because the operating condition of the SCO_2 heat exchanger is at a high temperature and high pressure, it is important to determine an appropriate material that can endure extreme operating conditions. Based on the operating condition, the maximum material surface temperature was assumed to be 650 °C. The maximum allowable stress values were considered as criteria for the heat exchanger material selection [23]. According to the maximum allowable stress values for candidate materials in the SCO_2 heat exchanger, S31042, S34709, and S34710 were selected because they have good corrosion and carburization resistance in SCO_2 environments. Based on the experimental results of the corrosion and carburization resistance, experience at similar operating conditions, cost, and availability, S34709 was finally selected as the SCO_2 heat exchanger material.

3. Design of Supercritical CO_2 Heat Exchanger

Figure 1 illustrates a schematic of the SCO_2 heat exchanger tube bundle. The tube specification was based on commercial availability and cost. The available tube diameter and the thickness of the S34709 material were 21.7 mm and 4.9 mm, respectively. The basic configuration of the SCO_2 heat exchanger had a rectangular duct fed by a flue gas. The heat transfer from the hot flue gas to the cold SCO_2 occurred in the rectangular duct. A staggered tube array with a counter-crossflow arrangement was considered. The tube length pitch (S_L) and tube height pitch (S_T) were 35 mm and 60 mm, respectively. The selection of the tube length pitch was based on the minimum thickness for pipe bends for induction and incrementing bending [24]. The straight line of the heat exchanger tube was 800 mm. The height and length of the tube were 471.7 mm and 2086.7 mm, respectively. Figure 2 shows the heat exchanger nozzle, header, and tube supporting structure. The total length of the SCO_2 heat exchanger was estimated as 5132 mm, and the lengths of the SCO_2 heat exchanger combustor and SCO_2 heat exchanger chamber were 1577 mm and 2765 mm, respectively. As a concept for the tube-supporting structure, rectangular plates made by the welding method were considered.

Preliminary simulation of the SCO_2 heat exchanger was performed using an in-house heat exchanger analysis code. Figure 3 shows a program flow chart for the analysis of the SCO_2 heat exchanger. An effective number of the transfer unit method was used for the simulation code. The thermo-properties of the flue gas and SCO_2 were obtained from the NIST REFPROP Database 23, Version 9.1 [25]. The fluid properties, such as Reynolds number, heat transfer coefficient, and friction factor, were calculated based on the thermal properties of the working fluids. Then, the heat transfer characteristics, such as overall heat transfer coefficient, number of transfer units, and effectiveness, were computed to obtain the outlet temperature of the working fluid. The pressure drop of each fluid was then obtained when the calculation of the outlet temperature was converged. For the shell side, the heat transfer correlation proposed by Zukauskas [26] was used. The pressure drop correlation for the staggered tube banks was employed [27]. For the tube side, the heat transfer correlation proposed by Gnielinski [28] was used, and the pressure drop was calculated by considering the entrance, momentum, core friction, and exit effects [29]. The validation of the developed code was verified with other commercially available heat exchanger code. The simulation results showed that the outlet temperature (600.6 °C) can be obtained with the present design considerations of the SCO_2 heat exchanger. In addition, the pressure drop on the SCO_2 side satisfied the design constraints (<1.5 bar). However, there is a limitation in using the in-house heat exchanger code because it only represents the outlet conditions. This means that it is difficult to find local heat transfer characteristics along the heat exchanger tubes as well as thermal stress along the tubes. Therefore, the analysis of thermal characteristics was conducted using commercial three-dimensional CFD codes.

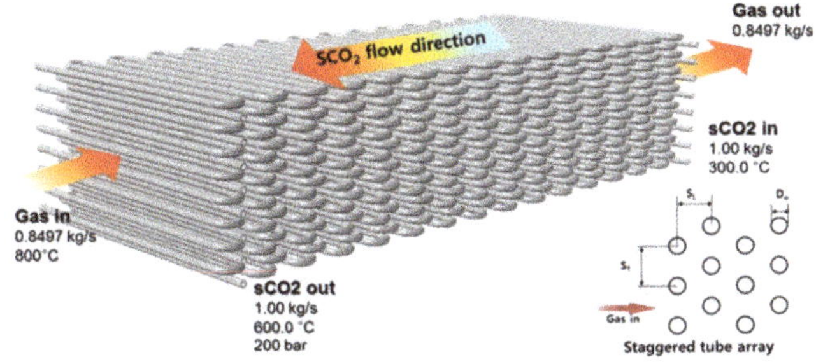

Figure 1. Schematic of the SCO_2 heat exchanger tube bundle.

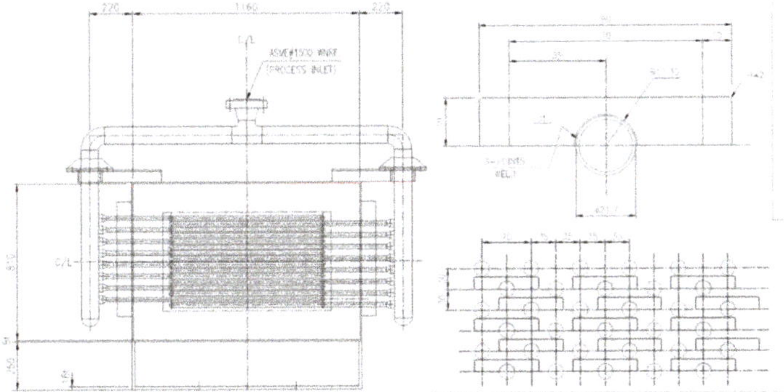

Figure 2. SCO_2 heat exchanger header, nozzle, and tube-supporting structure.

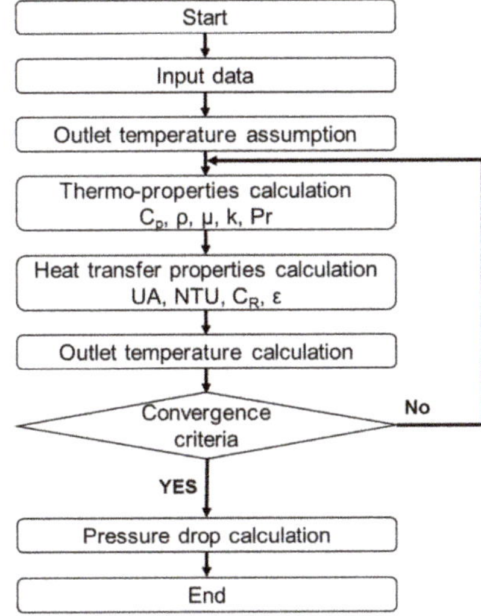

Figure 3. Heat exchanger performance analysis flow chart.

4. CFD Analysis of a Supercritical CO_2 Heat Exchanger

Three-dimensional commercially available CFD codes were used to analyze the flow and thermal stress characteristics of the SCO_2 heat exchanger. Flow and thermal stress analysis were performed.

Based on the temperature and heat transfer coefficient distributions obtained from the flow analysis, thermal stress analysis was conducted to evaluate the tube's integrity at the SCO_2 heat exchanger operating condition.

4.1. Flow Analysis

The thermal characteristics of the SCO_2 heat exchanger were analyzed using a commercial CFD code, CFX. The overall performance simulation of the SCO_2 heat exchanger was difficult due to the computing power. Therefore, a design of the scaled SCO_2 heat exchanger was created for the CFD analysis. Figure 4 shows the scaled SCO_2 heat exchanger: Figure 4a indicates the overall three-dimensional geometry, and Figure 4b illustrates the staggered tube array used in the flow CFD analysis. In the SCO_2 heat exchange chamber, three heat exchanger tubes with a staggered tube array were positioned. The center of the heat exchanger tube reflected the full-scale tube specification, while the half-scale tube specification was considered in the bottom and top of the tubes. The numbers of nodes and elements for the analysis were 9,555,606 and 10,698,900, respectively. Figure 5 shows the mesh distributions on the scaled SCO_2 heat exchanger. Figure 5a illustrates the overall mesh distributions in the scaled SCO_2 heat exchanger, Figure 5b indicates the mesh structure around the straight tube position, and Figure 5c shows the mesh formation near the tube bending location. Precise cells near the wall surface were considered to keep the turbulence effect. Several grid layers were applied around the heat exchanger tubes to consider the wall effect on the working fluids.

For the turbulence model, a two-equation turbulence model of RNG k-ε was used: k is the turbulence kinetic energy, which is defined by the fluctuation in velocity, and ε is the turbulence eddy dissipation. The RNG k-ε model was improved from the standard k-ε model, and it was derived using the renormalization group theory. The RNG k-ε model has an additional term, which considers the eddy dissipation, average shear stress, and swirl effect. These features provide a more accurate turbulence model compared to the standard k-ε model. The transport equations for turbulence generation and dissipation are the same as those for the standard k-ε model, but the model constants differ [30].

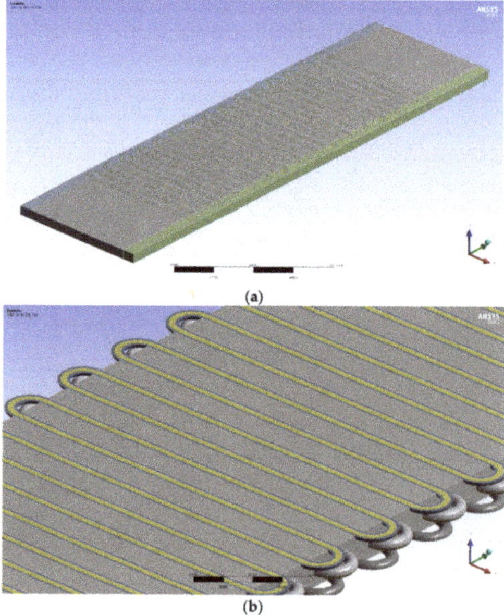

Figure 4. Scaled SCO_2 heat exchanger; (**a**) overall CFD analysis structure; (**b**) detailed view around heat exchanger tubes.

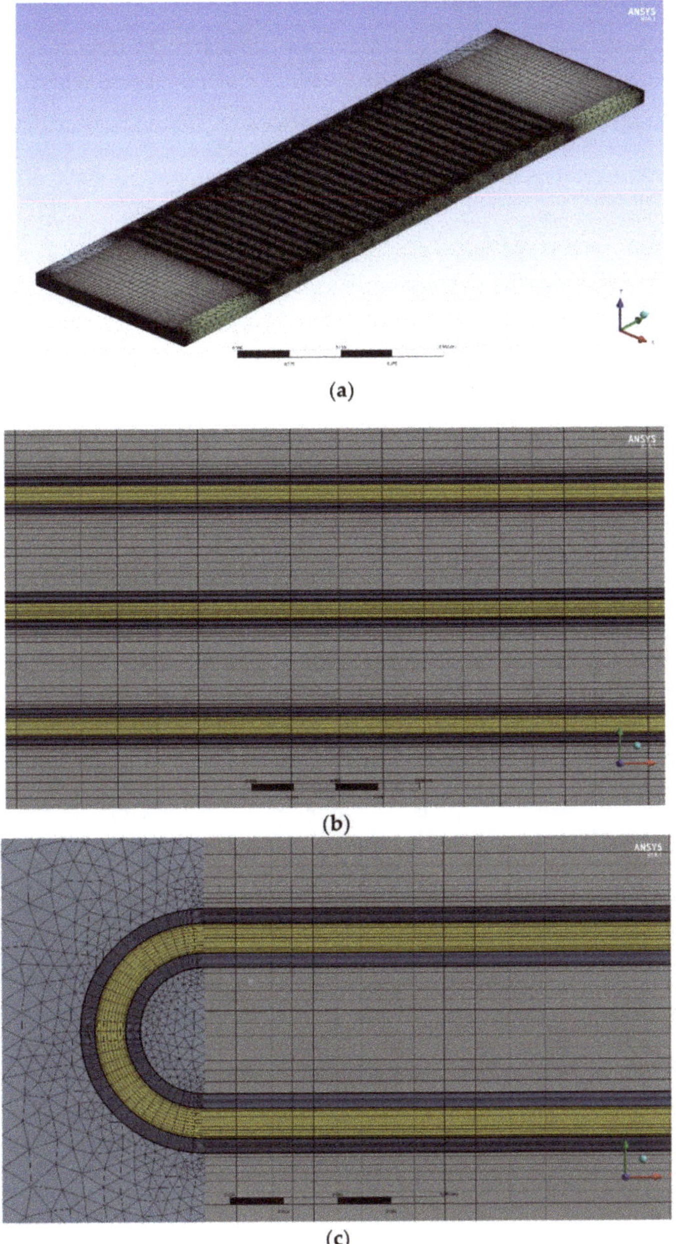

Figure 5. Mesh distributions on the scaled SCO_2 heat exchanger: (**a**) overall mesh distributions, (**b**) mesh distributions around the straight tube position, (**c**) mesh formation near the tube bending area.

For the continuity equation:

$$\frac{\partial \rho}{\partial t} + \frac{\partial}{\partial x_j}\left(\rho U_j\right) = 0 \tag{1}$$

For the momentum equation:

$$\frac{\partial \rho U_i}{\partial t} + \frac{\partial}{\partial x_j}\left(\rho U_i U_j\right) = -\frac{\partial p'}{\partial x_i} + \frac{\partial}{\partial x_j}\left(\mu_{eff}\left(\frac{\partial U_i}{\partial x_j} + \frac{\partial U_j}{\partial x_i}\right)\right) + S_M \tag{2}$$

For the transport equation for turbulence dissipation:

$$\frac{\partial(\rho\varepsilon)}{\partial t} + \frac{\partial}{\partial x_j}(\rho U_i \varepsilon) = \frac{\partial}{\partial x_j}\left[\left(\mu + \frac{\mu_t}{\sigma_{\varepsilon RNG}}\right)\frac{\partial \varepsilon}{\partial x_j}\right] + \frac{\varepsilon}{k}\left(C_{\varepsilon 1RNG}P_k - C_{\varepsilon 2RNG}\rho\varepsilon + C_{\varepsilon 1RNG}P_{\varepsilon b}\right) \tag{3}$$

where ρ is the density, p' is the modified pressure, μ is the viscosity, S_M is the sum of body forces and P_k is the shear production of turbulence, while $\sigma_{\varepsilon RNG}$, $C_{\varepsilon 1RNG\ constant}$, and $C_{\varepsilon 2RNG\ constant}$ are the RNG k-ε constants.

The boundary conditions were based on the actual operation conditions of the SCO_2 heat exchanger. For the actual SCO_2 heat exchanger, the number of heat exchanger tubes was 16, as shown in Figure 1. Uniform mass flow distributions on each heat exchanger tube were assumed: the mass flow rate of the SCO_2 in each tube was 0.0625 kg/s. For the flue gas in the scaled SCO_2 heat exchanger, the mass flow rate was calculated as 0.1062 kg/s. The fluid properties were implemented in the CFD code with a real gas properties table format, using NIST REFPROP Database 23, Version 9.1 [25]. For the convergence criteria, the SCO_2 outlet temperature was monitored in each step as well as solution imbalances (mass, momentum, and energy) less than 1%.

Figure 6 shows the CFD results for the center position of the SCO_2 heat exchanger: Figure 6a,b show the temperature and velocity distributions, respectively. High-velocity regions were focused on the left and right sides of the SCO_2 heat exchange chamber because there was empty space for installing the SCO_2 heat exchanger and tube displacement margin due to the thermal stress. However, the in-house code did not consider the empty space at the corner of the SCO_2 heat exchange chamber. The empty space could have resulted in different simulation results between the in-house code and the CFD analysis. However, the outlet temperature of the SCO_2 showed a similar performance (602 °C) compared to the in-house code result. The pressure drop of the SCO_2 flow path was calculated as 0.374 bar, which is lower than the design constraint. Further pressure drops should be considered in the heat exchanger headers and nozzles.

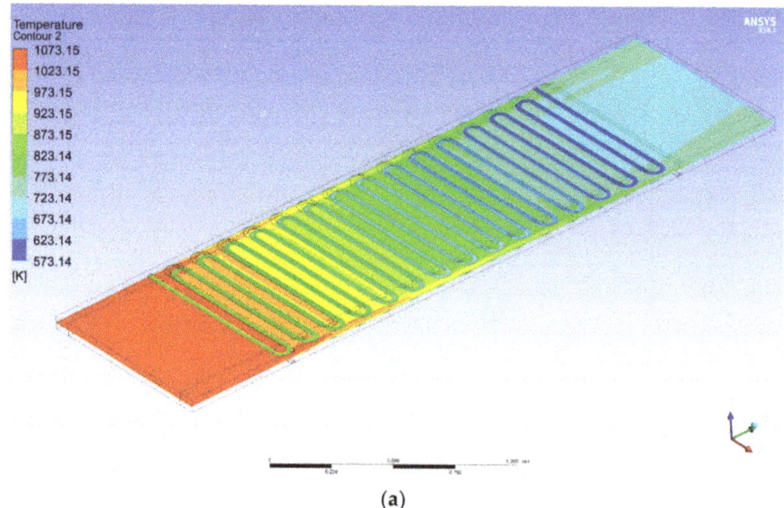

(a)

Figure 6. *Cont.*

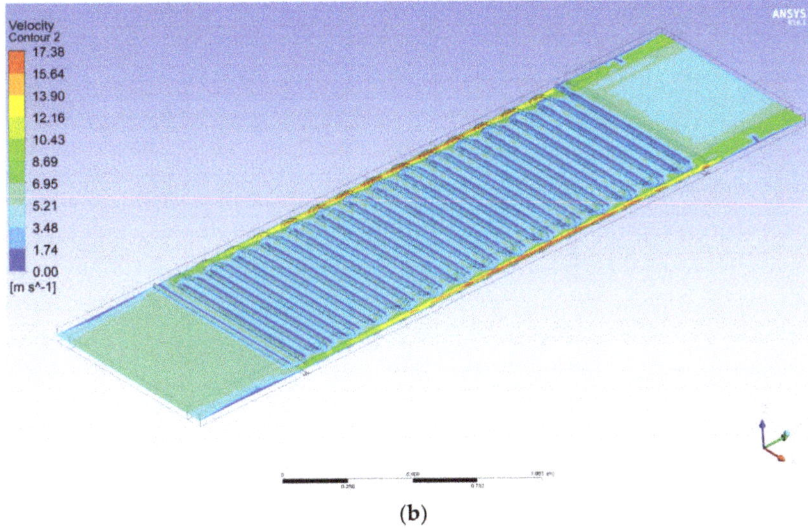

(b)

Figure 6. CFD results at the centerline of the scaled SCO_2 heat exchanger: (**a**) temperature and (**b**) velocity distributions.

The temperature and heat transfer coefficient distributions on the heat exchanger are important because these parameters can influence the tube's integrity. Figure 7 shows the temperature and heat transfer coefficient distributions on the inner and outer heat exchanger tubes. Figure 7a,b are the results of the tube's outer surface and Figure 7c,d are the results of the tube's inner surface. The tube's maximum inner and outer surface temperatures were 647 °C and 637 °C, respectively. The tube's maximum surface temperature was lower than the temperature assumption value (650 °C) in the heat exchanger material selection. For the tube's inner area, the maximum heat transfer coefficient was about 4000 W/m^{2}K, which was located at the tube's bending location. For the outer tube area, the maximum heat transfer coefficient was located near the outlet position of the flue gas and was 70 W/m^{2}K. Based on the temperature and heat transfer coefficient distributions on the heat exchanger tube, thermal stress analysis was performed.

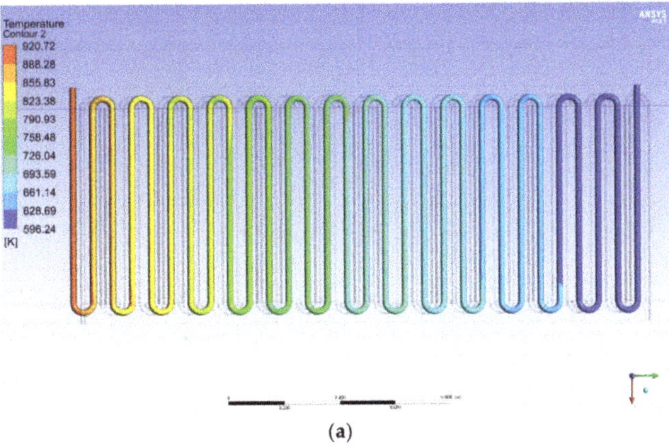

(a)

Figure 7. *Cont.*

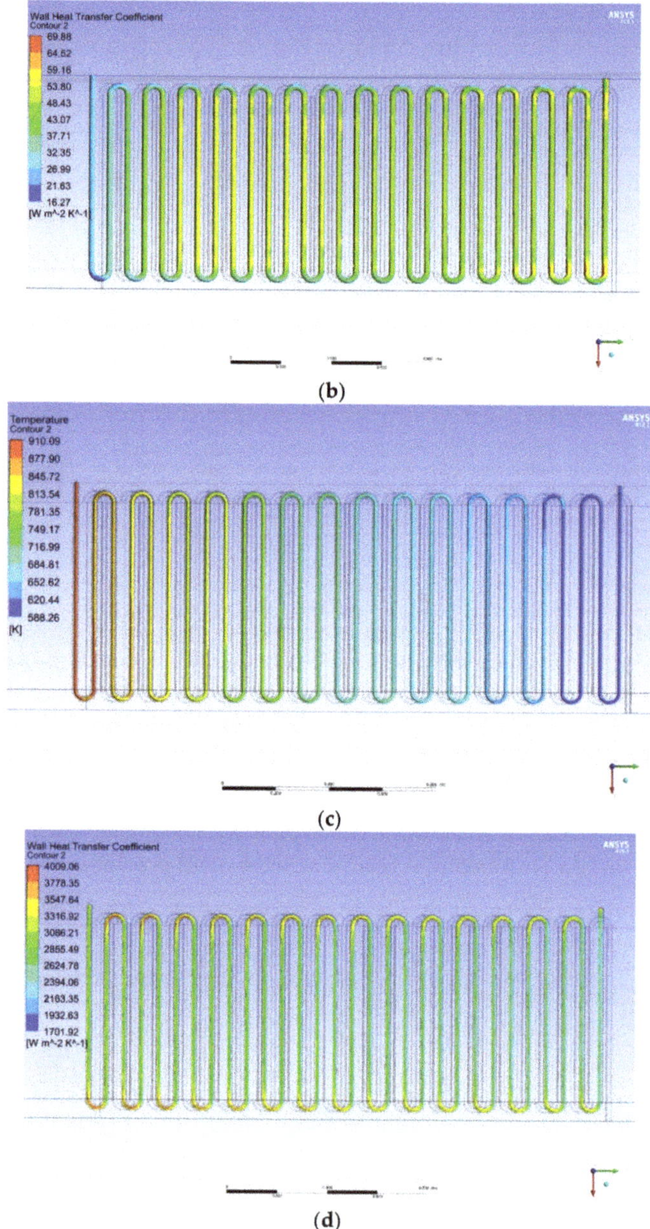

Figure 7. CFD results around the heat exchanger tube: (**a**) temperature distributions on the outer tube surface, (**b**) heat transfer coefficient distributions on the outer tube surface, (**c**) temperature distributions on the inner tube surface, (**d**) heat transfer coefficient distributions on the inner tube surface.

4.2. Thermal Stress Analysis

To show the integrity of the SCO_2 heat exchanger tube, thermal stress analysis was conducted using a commercial CFD code, ABAQUS [31]. High-temperature regions of the heat exchanger tube located near the SCO$_2$ outlet region (flue inlet region) were analyzed. Therefore, 10 straight lines with 10 bending flow paths were modeled for the thermal stress simulation. With consideration of the symmetrical structure of the heat exchanger tube, a half scale of the heat exchanger tube was analyzed. 1,114,135 elements were used for the thermo-mechanical analysis. Three-dimensional continuum element DCC3D8 was used to obtain the temperature distribution, and the stress analysis was carried out using the C3D8 continuum solid element.

In order to obtain temperature distributions for the heat exchanger tube, temperature and heat transfer coefficient profiles obtained from the thermal analysis were used as the input for thermal stress analysis. Because the pressure drop in the heat exchanger and heat exchange chamber was negligible compared to the operating condition, the pressures in the SCO$_2$ heat exchanger tube and the chamber were assumed to be 200 bar and atmospheric pressure, respectively. Figure 8 illustrates the temperature distribution in the scaled heat exchanger tube. A solid temperature of the heat exchanger tube is closer to the SCO$_2$ temperature because the SCO$_2$ heat transfer coefficient is higher than that of the flue gas. The evaluation of thermal stress analysis was based on the allowable stress value of S34709. The allowable stress value at a temperature of 650 °C is 539 bar, which is based on ASME Sec. II [23]. The finite element stress analysis was performed with three cases separately: (1) thermal loading, (2) pressure loading, and (3) thermal and pressure loading.

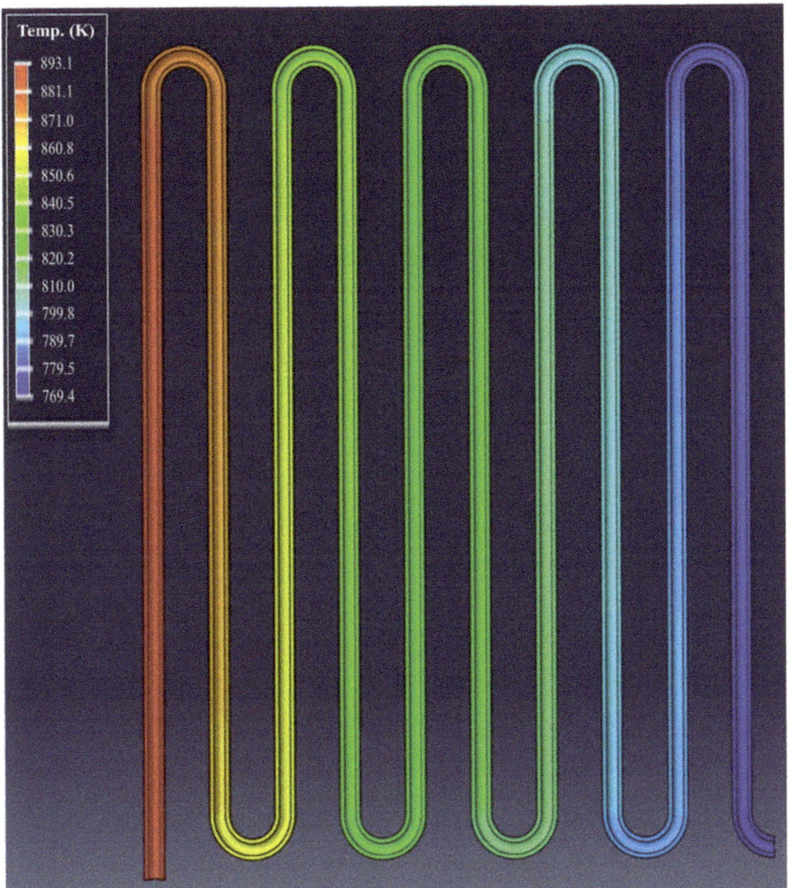

Figure 8. Heat exchanger tube temperature distributions used in the thermal stress analysis.

Figure 9 shows the tube displacement due to the temperature distribution in the heat exchanger tube. Figure 9a,b represent the tube displacement shape and the principal strain distributions, respectively. The high-temperature area near the exit of the SCO_2 had a large tube expansion, and it gradually decreased. The principal strain distributions showed similar distributions compared to the tube displacement. The maximum principal strain value was discovered to be 0.017. If the deformation of the heat exchanger tube is not considered, thermal buckling due to excessive pressure stress can occur. In the present SCO_2 heat exchanger design, empty space in the chamber was considered, which maintained the integrity of the tube even when the tube displacement appeared.

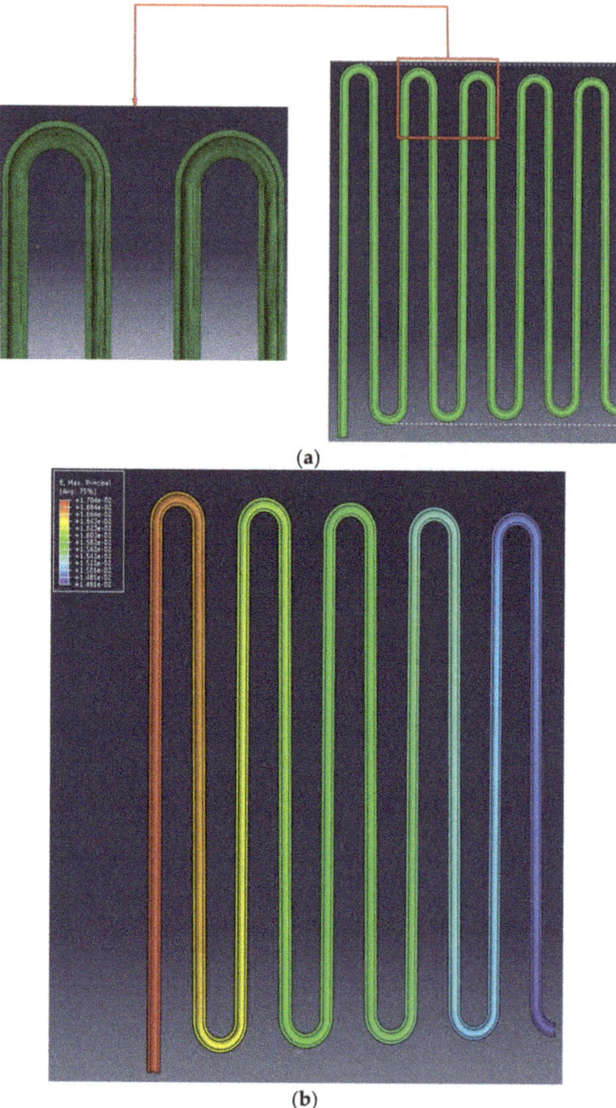

Figure 9. Tube displacements on the tube: (**a**) tube displacement shape, (**b**) the maximum principal strain distributions.

Von-Mises stress distributions are shown in Figure 10 along the scaled heat exchanger tube. Figure 10a,b are the results of the thermal and pressure loading cases, respectively. In the case of thermal loading, the maximum stress was found near the tube bending area. On the other hand, the maximum stress was located at the tube bending area for the pressure loading condition. The maximum local von-Mises stress values were 35.5 bar and 516 bar for the thermal and pressure loading cases, respectively. It was confirmed that these stress values were lower than the allowable stress value. Figure 10c is the result of thermal stress analysis considering both the thermal and pressure loading cases. The stress distributions in the tube were similar to the combination of the thermal and pressure loading cases. The maximum von-Mises value was calculated as 523 bar, which is lower than the allowable stress value considered in the present study. The stress evaluation loaded on the heat exchanger tube was conducted based on the ASME Sec. VIII. The membrane stress of the heat exchanger tube was 287 bar, while the allowable stress value was 539 bar. The sum of the membrane stress and the bending stress was 378 bar, which is lower than the constraint value ($1.5 \times$ allowable stress value = 808 bar). Therefore, the stress state of the SCO_2 heat exchanger satisfied the ASME criteria.

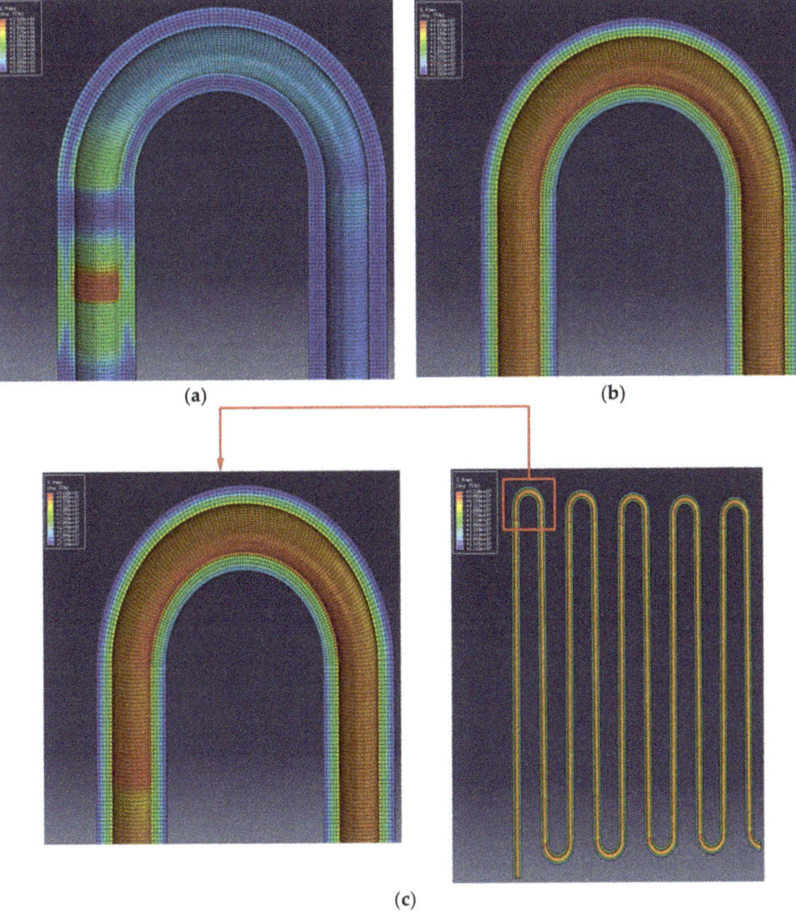

Figure 10. Von-Mises stress distributions: (**a**) thermal loading, (**b**) pressure loading, (**c**) thermal and pressure loading cases.

5. Conclusions

This study focused on the design of an SCO_2 heat exchanger for obtaining high-temperature and high-pressure heat exchange technologies under an SCO_2 environment. A tubular type heat exchanger was selected because it has high durability in extreme conditions, such as having low pressure losses in both the hot and cold sides. The heat exchanger material selection was conducted based on the maximum allowable stress, corrosion resistance, cost, and availability. A staggered tube array with a counter-cross flow arrangement was determined and the overall size of the SCO_2 heat exchanger was based on the tube bending criteria and the results of in-house heat exchanger performance code. Commercially available three-dimensional CFD codes were then used to analyze the flow and thermal characteristics of the SCO_2 heat exchanger. The temperature and heat transfer coefficient distributions on the SCO_2 heat exchanger were analyzed. Then, thermal stress analysis was conducted based on the obtained flow analysis results. The stressed state of the SCO_2 heat exchanger was evaluated based on the ASME procedure. The membrane stress, bending stress, and local stress were lower than the allowable stress. The results indicate that the stress of the present heat exchanger satisfied the ASME criteria. Based on the design of the SCO_2 heat exchanger, the manufacturing process can be performed.

Author Contributions: Conceptualization, H.S. and J.E.C.; Methodology, H.S. and J.E.C.; Software, H.S., J.K., and Y.-W.K.; Validation, H.S., J.K., and Y.-W.K.; Formal analysis, H.S., J.E.C., and I.S.; Investigation, H.S., J.K., and I.S.; Resources, H.S., Data Curation, H.S., J.K., and Y.-W.K.; Writing-Original Draft, H.S.; Writing-Review & Editing, J.E.C.; Visualization, H.S.; Supervision, H.S.; Project administration, H.S.; Funding acquisition, J.E.C. All authors have read and agreed to the published version of the manuscript.

Funding: This work was financially supported by the institute of Civil Military Technology Cooperation funded by the Defense Acquisition Program Administration and Ministry of Trade, Industry and Energy of Korean government under grant No. 17-CM-EN-04.

Conflicts of Interest: The authors declare no conflict of interest.

Nomenclature

Nomenclature	
S	pitch
k	turbulence kinetic energy
P	pressure
T	temperature
P	shear production
U	velocity
Greek symbols	
ε	turbulence eddy dissipation
ρ	density
μ	viscosity
Subscripts	
c	critical
L	length
T	height

References

1. Feher, E.G. The Supercritical Thermodynamic Power Cycle. In Proceedings of the Intersociety Energy Conversion Engineering Conference, Miami Beach, FL, USA, 13–17 August 1967. Douglas Paper No. 4348.
2. Angelino, G. Real Gas Effects in Carbon Dioxide Cycles. In Proceedings of the ASME 1969 Gas Turbine Conference and Products Show, Paper No. 69-GT-103. Cleveland, OH, USA, 9–13 March 1969.
3. Dostal, V.; Driscoll, M.J.; Hejzlar, P. *A Supercritical Carbon Dioxide Cycle for Next Generation Nuclear Reactors*; MIT-ANP-TR-100; Massachusetts Institute of Technology: Cambridge, MA, USA, 2004.

4. Wright, S.A.; Fuller, R.; Pickard, P.S.; Vernon, M.E. Initial status and test results from a supercritical CO_2 Brayton cycle test loop. In Proceedings of the International Conference on Advances in Nuclear Power Plants, Anaheim, CA, USA, 8–12 June 2008.

5. Wright, S.A.; Pickard, P.S.; Fuller, R.; Radel, R.F.; Vemon, M.E. Supercritical CO_2 Brayton cycle power generation development program and initial results. In Proceedings of the ASME Power Conference, Albuquerque, NM, USA, 21–23 July 2009.

6. Wright, S.A.; Radel, R.F.; Vernon, M.E.; Rochau, G.E.; Pickard, P.S. *Operation and Analysis of a Supercritical CO_2 Brayton Cycle*; SAND2010-0171; Sandia National Laboratories: California, CA, USA, 2010.

7. Clementoni, E.M.; Cox, T.L.; Sprague, C.P. Startup and operation of a supercritical carbon dioxide Brayton cycle. *ASME J. Eng. Gas Turbines Power* **2014**, *136*, 071701. [CrossRef]

8. Clementoni, E.M.; Cox, T.L.; King, M.A. Off-nominal component performance in a supercritical carbon dioxide Brayton cycle. *ASME J. Eng. Gas Turbines and Power* **2016**, *138*, 011703. [CrossRef]

9. Clementoni, E.M.; Cox, T.L.; King, M.A. *Steady-State Power Operation of a Supercritical Carbon Dioxide Brayton Cycle*; ASME Turbo Expo: Seoul, Korea, 2016.

10. Held, T.J. Initial test results of a megawatt-class supercritical CO_2 heat engine. In Proceedings of the 4th International Supercritical CO_2 Power Cycles Symposium, San Antonio, TX, USA, 9–10 September 2014.

11. Held, T.J. Commercialization of Supercritical CO_2 Power Cycles. In Proceedings of the 1st European Seminar on Supercritical CO_2 Power Systems, Vienna, Austria, 29–30 September 2016.

12. Moore, J.; Brun, K.; Evans, N.; Kalra, C. Development of 1 MWe supercritical CO_2 test loop. In Proceedings of the ASME Turbo Expo, Montreal, QC, Canada, 15–19 June 2015.

13. Hoopes, K.; Rimpel, A. Development of a 2.6MW heater for SUNSHOT sCO_2 turbine mechanical test. In Proceedings of the 5th International Supercritical CO_2 Power Cycles Symposium, San Antonio, TX, USA, 28–31 March 2016.

14. Moore, J.; Cich, S.; Day, M.; Allison, T.; Wade, J.; Hofer, D. Commissioning of a 1 MWe supercritical CO_2 test loop. In Proceedings of the 6th International Supercritical CO_2 Power Cycles Symposium, Sacramento, CA, USA, 27–29 March 2018.

15. Department of Energy (DOE). Energy Department Announces New Investments in Supercritical Transformational Electric Power Program. Available online: https://www.energy.gov/ne/articles/energy-department-announces-new-investments-supercritical-transformational-electric#:~{}:text=WASHINGTON%20%2D%2D%20The%20U.S.%20Department,electrical)%20Supercritical%20Carbon%20Dioxide%20 (accessed on 15 December 2015).

16. Bush, V. GTI STEP forward on sCO_2 power. In Proceedings of the 6th International Supercritical CO_2 Power Cycles Symposium, Sacramento, CA, USA, 27–29 March 2018.

17. Huang, M.; Tang, C.; McClung, A. Steady state and transient modeling for the 10 MWe SCO_2 test facility program. In Proceedings of the 6th International Supercritical CO_2 Power Cycles Symposium, Sacramento, CA, USA, 27–29 March 2018.

18. Cha, J.E.; Ahn, Y.; Lee, J.; Lee, J.I.; Choi, H.L. Installation of the Supercritical CO_2 Compressor Performance Test Loop as a First Phase of the SCIEL facility. In Proceedings of the 4th International Supercritical CO_2 Power Cycles Symposium, Pittsburgh, PA, USA, 9–10 September 2014.

19. Cha, J.E.; Bae, S.W.; Lee, J.; Cho, S.K.; Lee, J.I.; Park, J.H. Operation Results of a Closed Supercritical CO_2 Simple Brayton Cycle. In Proceedings of the 5th International Supercritical CO_2 Power Cycles Symposium, Chicago, IL, USA, 28–31 March 2016.

20. Cho, J.; Shin, H.; Ra, H.; Lee, G.; Rho, C.; Lee, B.; Baik, Y. Research on the development of a small-scale supercritical carbon dioxide power cycle experimental test loop. In Proceedings of the 5th International Supercritical CO_2 Power Cycles Symposium, Chicago, IL, USA, 28–31 March 2016.

21. Cho, J.; Shin, H.; Cho, J.; Ra, H.; Roh, C.; Lee, B.; Lee, G.; Choi, B.; Baik, Y. Preliminary power generating operation of the supercritical carbon dioxide power cycle experimental test loop with a turbo-generator. In Proceedings of the 6th International Supercritical CO_2 Power Cycles Symposium, Sacramento, CA, USA, 27–29 March 2018.

22. Natesan, K.; Moisseytsev, A.; Majumdar, S. Preliminary Issues Associated with the Next Generation Nuclear Plant Intermediate Heat Exchanger Design. *J. Nucl. Mater.* **2009**, *392*, 307–315. [CrossRef]

23. ASME Boiler and Pressure Vessel Committee on Materials. *ASME Boiler and Pressure Vessel Code Section II*; American Society of Mechanical Engineers (ASME): New York, NY, USA, 2015.

24. ASME Boiler and Pressure Vessel Committee on Materials. *ASME Boiler and Pressure Vessel Code Section III*; American Society of Mechanical Engineers (ASME): New York, NY, USA, 2015.

25. Lemmon, E.W.; Huber, M.L.; McLinden, M.O. *NIST Standard Reference Database 23: Reference Fluid Thermodynamic and Transport Properties-REFPROP, Version 9.1*; National Institute of Standards and Technology: Gaithersburg, DC, USA, 2013.

26. Zukauskas, A.A. *High-Performance Single-Phase Heat Exchangers*; Hemisphere Publishing Corp.: New York, NY, USA, 1989.

27. Zukauskas, A.A.; Makarevicius, V.J.; Slanciauskas, A.A. *Heat Transfer in Banks of Tubes in Crossflow of Fluid; Thermophysics 1*; Mintis: Vilnius, Lithuania, 1968; pp. 47–68.

28. Gnielinski, V. New Equations for Heat and Mass Transfer in Turbulent Pipe and Channel Flow. *Int. Chem. Eng.* **1976**, *16*, 359–368.

29. Shah, R.K.; Sekulic, D.P. *Fundamentals of Heat Exchanger Design*; John Wiley and Sons, Inc.: Hoboken, NJ, USA, 2003.

30. ANSYS. *ANSYS CFX-Solver Theory Guide*; ANSYS, Inc.: Canonsburg, PA, USA, 2017.

31. ABAQUS/CAE. *User's Guide Version*; Dassault Systems Simulia Corp.: Providence, RI, USA, 2013.

Article

Performance Analysis of the Supercritical Carbon Dioxide Re-compression Brayton Cycle

Mohammad Saad Salim, Muhammad Saeed [†] and Man-Hoe Kim [*]

School of Mechanical Engineering & IEDT, Kyungpook National University, Daegu 41566, Korea; msaadsalim@hotmail.com (M.S.S.); saeed.aarib@gmail.com (M.S.)

* Correspondence: kimmh@asme.org; Tel.: +82-53-950-5576
† Present Address: Mechanical Engineering Department, Khalifa University of Science and Technology, SAN Campus, P.O. Box 2533, Abu Dhabi, UAE.

Received: 30 December 2019; Accepted: 5 February 2020; Published: 7 February 2020

Abstract: This paper presents performance analysis results on supercritical carbon dioxide (sCO_2) re-compression Brayton cycle. Monthly exergy destruction analysis was conducted to find the effects of different ambient and water temperatures on the performance of the system. The results reveal that the gas cooler is the major source of exergy destruction in the system. The total exergy destruction has the lowest value of 390.1 kW when the compressor inlet temperature is near the critical point (at 35 °C) and the compressor outlet pressure is comparatively low (24 MPa). The optimum mass fraction (x) and efficiency of the cycle increase with turbine inlet temperature. The highest efficiency of 49% is obtained at the mass fraction of $x = 0.74$ and turbine inlet temperature of 700 °C. For predicting the cost of the system, the total heat transfer area coefficient (UA_{Total}) and size parameter (SP) are used. The UA_{Total} value has the maximum for the split mass fraction of 0.74 corresponding to the maximum value of thermal efficiency. The SP value for the turbine is 0.212 dm at the turbine inlet temperature of 700 °C and it increases with increasing turbine inlet temperature. However the SP values of the main compressor and re-compressor increase with increasing compressor inlet temperature.

Keywords: re-compression Brayton cycle; carbon dioxide; supercritical; thermodynamic; exergy; cycle simulation; design point analysis

---

1. Introduction

The main cause of pollution is the combustion of fossil fuels to create energy for heavy industrialization and urbanization. Fossil fuel reserves are diminishing due to this process; thus, a big demand for power generation from green energy sources at high efficiency has been created. Global warming is another big concern, as has been pointed out by the United Nations Framework Convention on Climate Change (UNFCC). The proposal from the conference was to undertake efforts so that the rise in global average temperature increase could be limited to well below 2 °C above preindustrial levels. Due to the use of fossil fuels and harmful working fluids, significantly harmful effects on the environment are causing problems such as global warming and acid rain. The effects of pollution tend to bring unpredictable changes in the global climate, as has been asserted by the Intergovernmental Panel on Climate Change (IPCC), and rising sea levels are making large parts of the Earth uninhabitable [1]. Thus, green sources of energy are the need of the hour to solve these issues, which has led to research being conducted on different forms of green energy, such as biogas [2,3], geothermal energy [4,5], energy from human excreta [6] and solar energy [7]. In order to cope with the aforementioned global climate challenges, carbon dioxide (CO_2)-based power systems present an environmentally friendly option and are capable of providing power at high efficiency.

The Rankine cycle and the air-standard Brayton cycle are well-known thermodynamic cycles. The benefit of the Rankine cycle is that high efficiency can be achieved because the pump consumes a very small amount of work, since compression is carried out when the working fluid is in the liquid state [8]. The advantage of the Brayton cycle is that the turbine inlet temperature is high; thus, it can achieve high efficiency, but the disadvantage is that the work consumed by the compressor is very large. Due to this, the air-standard Brayton cycle's efficiency is not significantly higher than that of the steam-based Rankine cycle. The primary advantage of the supercritical CO_2 (sCO_2) Brayton cycle is that the positive points of the steam-based Rankine cycle and air-standard Brayton cycle are both combined. The turbine inlet temperature in the sCO_2 Brayton cycle is high. Moreover, since the compressor operates near the CO_2 critical point at very high pressure at which the density is significantly high and the compressibility factor is small, the work that is consumed by the compressor is significantly low. The sCO_2 Brayton cycle operates above the critical point, so the need for condensing the system is removed and the system has a simple layout. Since the sCO_2 re-compression Brayton cycle has very high operating pressures compared to the steam-based Rankine cycle, the size of the sCO_2-based power system's components is considerably smaller [9].

Currently, extensive research is being conducted on sCO_2-based power systems, and the sCO_2 Brayton cycle can be found in a variety of arrangements in the literature, such as in reheated and intercooled re-compression layouts [10]. Crespi et al. [11] reviewed the different single and combined layouts of sCO_2 Brayton cycle power systems with efficiencies of 40%–50% and 50%–60%, respectively, while Saeed and Kim [8] analyzed a re-compression sCO_2 Brayton cycle power system with an integrated turbine design and optimization algorithm. In their study, they proposed that the cycle performs best when the inlet temperature of the compressor is set near the CO_2 critical temperature (i.e., 32–37 °C) and the compressor inlet pressure is set slightly above the CO_2 critical pressure (i.e., 7.8–8.1 MPa) along with a moderate pressure ratio (i.e., 2.9–3.1). Saeed et al. [12] carried out a design optimization and performance analysis of the sCO_2 re-compression Brayton cycle. They developed detailed mathematical models of the cycle components and simulation codes for the turbine, compressor and heat exchanger. These codes were used to analyze the performance of the cycle under the design conditions as well as off-design conditions.

To analyze the performances of thermal power systems, researchers have used different performance parameters, including thermal efficiency (η_{th}) and exergy efficiency (η_{ex}). Moreover, to indicate the cost and size of the thermodynamic system, parameters such as the total heat transfer area coefficient (UA_{Total}) and size parameter (SP_{Total}) have also been used to indicate the heat exchanger and turbomachinery sizes, respectively [13–15]. Patel et al. [16] studied the optimization of a waste-heat-based organic Rankine cycle (ORC)-powered cascaded vapor compression-absorption refrigeration system. They used the log mean temperature difference (LMTD) to determine the UA value of each heat exchanger in the system. The purpose was to minimize the UA value in order to minimize the area required for heat exchange, and thus minimize the cost of the heat exchangers in the system.

This paper presents the various benefits of sCO_2 power systems. A detailed investigation has been conducted for the performance of the system with regard to key performance parameters such as η_{th}, η_{ex}, UA and size parameter (SP) of the turbomachinery. The system analysis also has been performed on the basis of the changing ambient temperature (T_0) and water temperature ($T_{w,in}$) values to signify how the system's exergetic performance changed on a monthly basis. To the author's best knowledge, monthly/seasonal analysis using for sCO_2-based power systems are not available in the literature. Moreover, the performance of the system at different turbine inlet temperatures (T_7), compressor inlet temperatures (T_1) and compressor outlet pressures (P_2) is also presented to signify the effect of each variable on various performance parameters of the system, and to indicate which values are best for use as the cycle's design points.

2. Methodology

2.1. Cycle Processes

Figure 1 shows a cycle layout and temperature-entropy (T-s) diagram of the system considered in the study. As shown in Figure 1a, the system consists of a turbine, a primary heat exchanger, high and low-temperature recuperators, the main compressor, a re-compression compressor and a gas cooler. In comparison with the recuperated cycle, the re-compression Brayton cycle includes an added intermediate compressor, an additional recuperator and split/mixing flow values. A fraction of mass is taken from the mainstream and fed to the re-compression compressor bypassing the low-temperature recuperator (LTR). This fraction of flow enters the mainstream again increases its temperature before entering into the high-temperature recuperator (HTR). This arrangement in turns increases the thermal efficiency of the cycle (42% to 50% for cycles operating with the lowest and highest temperatures, 37 °C and 700 °C respectively [10]) by reducing heat rejection in the pre-cooler.

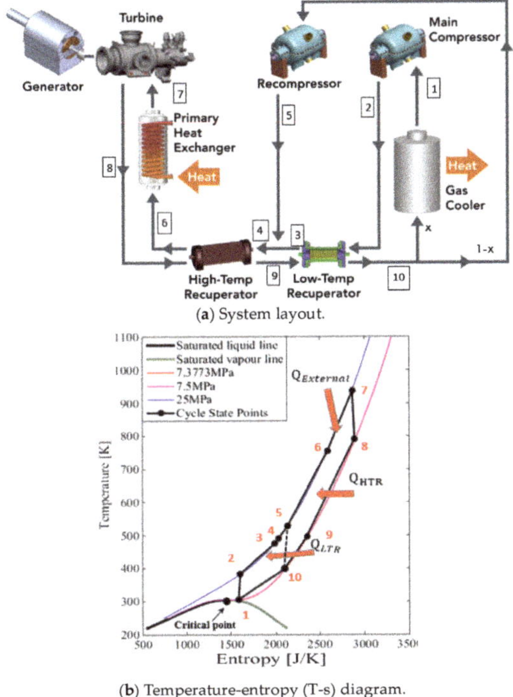

(a) System layout.

(b) Temperature-entropy (T-s) diagram.

Figure 1. Schematic of re-compression sCO_2 Brayton cycle.

The stream exiting the re-compressor in State 5 and the stream exiting the recuperator in State 3 are combined in State 4. This stream then passes the HTR where its temperature further increases to State 6. After this, the heat addition process takes place at the main heat exchanger and the stream reaches T_7 in State 7. In the turbine, the expansion process takes place until State 8. From State 8 to State 9, the HTR recuperates heat, and then, from State 9 to State 10, the LTR recuperates heat. After this, cooling takes place in the gas-cooler in State Point 1. The gas cooling process from State 10 to State 1 is used to transfer heat to the coolant (water) which is input to the gas cooler at the monthly temperature value (see Figure 2) and has to be heated to 40 °C for domestic uses, such as floor heating.

2.2. Design Parameters

The cycle simulations were developed using MATLAB [17], and the thermodynamic properties of CO_2 at each state point in the cycle were calculated NIST's REFPROP [18]. The design parameters and governing equations for defining the operation of the cycle are listed in Table 1.

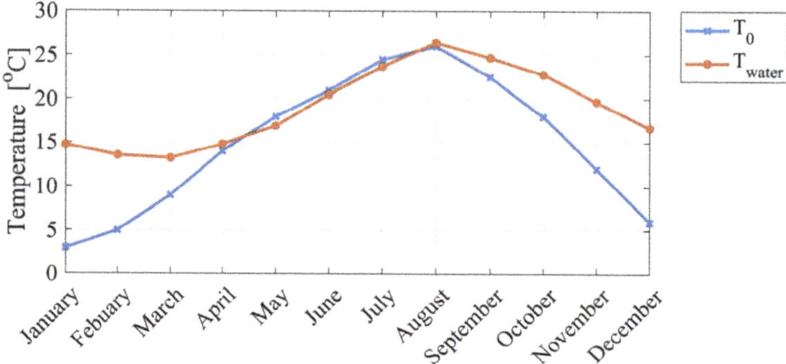

Figure 2. Ambient and water temperature data for the city of Busan [19].

Table 1. Design parameters used in sCO_2 cycle simulations.

Parameter	Symbol	Value
CO_2 mass flow rate, kg/s	m_{CO_2}	10
Main compressor inlet temperature, °C	T_1	35
Main compressor inlet pressure, MPa	P_1	7.8
Main compressor outlet pressure, MPa	P_2	24
Main compressor isentropic efficiency	η_{MC}	0.85
Re-compressor isentropic efficiency	η_{RC}	0.85
Turbine isentropic efficiency	η_T	0.93
Effectiveness of the HTR	ε_{HTR}	0.90
Effectiveness of the LTR	ε_{LTR}	0.90
Ambient temperature (other than seasonal exergy analysis), °C	T_0	24.5
Ambient pressure, MPa	P_0	0.1
Water inlet temperature in the gas cooler (other than seasonal exergy analysis), °C	$T_{w,in}$	23.7
Water outlet temperature in the gas cooler, °C	$T_{w,out}$	40

For the monthly exergy analysis presented in this study, ambient temperature (T_0) and water (coolant) inlet temperatures ($T_{w,in}$) in the gas cooler were used from the data for the city of Busan [19] as shown in Figure 2.

2.3. Energy and Exergy Analysis

To simplify the analysis, the following assumptions are made:

1. Pressure drops in the heat exchangers and pipes are neglected.
2. Steady-state operation is assumed for all the devices in the system.
3. Heat loss from the components is negligible.

The following set of governing equations represent the energy analysis for the sCO_2 cycle:

$$W_T = m_{CO_2}(h_7 - h_8),\tag{1}$$

$$W_{MC} = m_{CO_2}x(h_2 - h_1),\tag{2}$$

$$W_{RC} = m_{CO_2}(1 - x)(h_5 - h_{10}),\tag{3}$$

$$Q_{in} = m_{CO_2}(h_7 - h_6), \tag{4}$$

$$Q_{GC} = m_{CO_2}x(h_{10} - h_1), \tag{5}$$

$$W_{net} = W_T - W_{MC} - W_{RC}, \tag{6}$$

$$\eta_{th} = \frac{W_{net}}{Q_{in}}. \tag{7}$$

Thermal efficiency (η_{th}) is a parameter that takes into account how much of the energy input from the heat source is converted into the net shaft work by the turbine, but it does not reflect the irreversibilities that are involved in the sCO_2 re-compression Brayton cycle. The exergy analysis of the system is significantly useful, since energy is conserved but exergy is destroyed by the irreversibility [20]. The second law (exergy) efficiency of the system is an indicator of the maximum theoretical work that can be generated as the system is brought to equilibrium with the environment.

The specific exergy of flow at any state in the system is given by

$$e = h - h_0 - T_0(s - s_0). \tag{8}$$

By applying exergy balance over each component, the exergy destruction in the different components of the system can be defined as follows.

Primary heat exchanger:

$$I_{HX} = m_{CO_2}(e_6 - e_7) + m_s(e_{si} - e_{so}), \tag{9}$$

where m_s is the mass flow rate of the heat transfer fluid.

Main compressor:

$$I_{MC} = m_{CO_2}x(e_1 - e_2) + W_{MC}. \tag{10}$$

Re-compressor:

$$I_{RC} = m_{CO_2}(1 - x)(e_{10} - e_5) + W_{RC}. \tag{11}$$

Gas cooler:

$$I_{GC} = m_{CO_2}x(e_{10} - e_1) + m_w(e_{win} - e_{wout}). \tag{12}$$

Turbine:

$$I_T = m_{CO_2}(e_7 - e_8) - W_T. \tag{13}$$

High-temperature recuperator:

$$I_{HTR} = m_{CO_2}(e_8 - e_9 + e_4 - e_6). \tag{14}$$

Low-temperature recuperator:

$$I_{LTR} = m_{CO_2}[e_9 - e_{10} + x(e_2 - e_3)]. \tag{15}$$

The total exergy destruction in the system is given by

$$I_{Total} = I_{HX} + I_{MC} + I_{RC} + I_{GC} + I_T + I_{HTR} + I_{LTR}. \tag{16}$$

The heat rejection from the system in the gas cooling process, which can be utilized for district heating purposes, is determined from the following equation:

$$Q_{out} = m_w(h_{wout} - h_{win}). \tag{17}$$

By balancing the exergy throughout the whole system, the exergy input to the system is

$$E_{in} = W_{net} + I_{Total} + \left(\frac{T_0}{T_{w\,avg}} - 1\right)Q_{out}.$$

(18)

The exergy efficiency of the system is given by

$$\eta_{ex} = \frac{W_{net}}{E_{in}}.$$

(19)

The total irreversibility ratio (IR) of the system can be written as

$$IR = \frac{I_{total}}{E_{in}}.$$

(20)

2.4. Total Heat Transfer Area Coefficient and Total Size Parameter Value (SP_{Total})

UA_{Total} and SP_{Total} are useful parameters for estimating the cost of the system. UA_{Total} indicates the total area required for the heat exchangers in the sCO_2 re-compression Brayton cycle, and thus the cost associated with the investment and maintenance of the heat exchangers. The hypothesis is that differences between the heat transfer coefficients in the sCO_2 cycle heat exchangers are not significant. The total heat transfer area in the heat exchangers increases as a result of increasing UA_{Total}, thereby increasing the investment and maintenance costs of the heat exchangers, and thus the system [21]. A smaller value of UA_{Total} is desirable for a cost-effective design.

For a heat exchanger in general, we have

$$UA = \frac{Q}{\Delta T_M},$$

(21)

$$\Delta T_M = \frac{\Delta T_{max} - \Delta T_{min}}{\ln\left(\frac{\Delta T_{max}}{\Delta T_{min}}\right)},$$

(22)

where ΔT_M is logarithmic mean temperature difference and ΔT_{max} and ΔT_{min} are the maximum and minimum temperature differences, respectively, at the two ends of the heat exchanger. However, in the case of supercritical carbon dioxide Brayton cycle (sCO_2-BC), properties of the working fluid change swiftly, and the definition of the ΔT_M is not applicable in this case. In order to cope with the situation, the length of the heat exchanger was divided into N number of segments. The length of each segment was kept small enough that the variation within a particular small segment can be ignored, as shown in Figure 3. The value of the number of segments (N) depends upon the operation region of the heat exchanger. For the pre-cooler and LTR, the numbers of segments will be large, as these two heat exchangers operate close to the critical point, and properties of sCO_2 change at high rates. For HTR, in contract with LTR and pre-cooler, but operating away from the critical point where variation in the properties is small, a lesser number of segments would be required. Further details on the model can be found in the previous studies [8,12,22,23].

Based on the discretized model definition of the logarithmic mean temperature difference $(\Delta T_M)^i$, the ith segment is given by the Equation (23)

$$
\begin{aligned}
(\theta_2)^i &= \left(T^h\right)^i - (T^c)^i; \\
(\theta_1)^i &= \left(T^h\right)^{i+1} - (T^c)^{i+1}; \\
(\Delta T_M)^i &= \frac{\max\left((\theta_1)^i,(\theta_2)^i\right) - \min\left((\theta_1)^i,(\theta_2)^i\right)}{\log\left(\frac{\max\left((\theta_1)^i,(\theta_2)^i\right)}{\min\left((\theta_1)^i,(\theta_2)^i\right)}\right)}
\end{aligned}
$$

(23)

Thus, UA value through the ith segment could be computed as given by the following equation.

$$(UA)^i = \frac{(dq)i}{(\Delta T_M)^i}. \tag{24}$$

And UA value associated with the whole heat exchanger can be calculated using the following relation.

$$UA = \sum_{i=1}^{N} (UA)^i. \tag{25}$$

Later, combining the UA values for the sCO_2 system, we obtain

$$UA_{Total} = UA_{HX} + UA_{HTR} + UA_{LTR} + UA_{GC}. \tag{26}$$

SP indicates the size of the turbomachinery in the system, and a high SP_{Total} means comparatively large sizes of the turbine and compressors. Consequently, the investment and maintenance costs of the turbomachinery will increase, which, in turn, will increase the total cost of the system. Because of this, a small SP_{Total} is desirable.

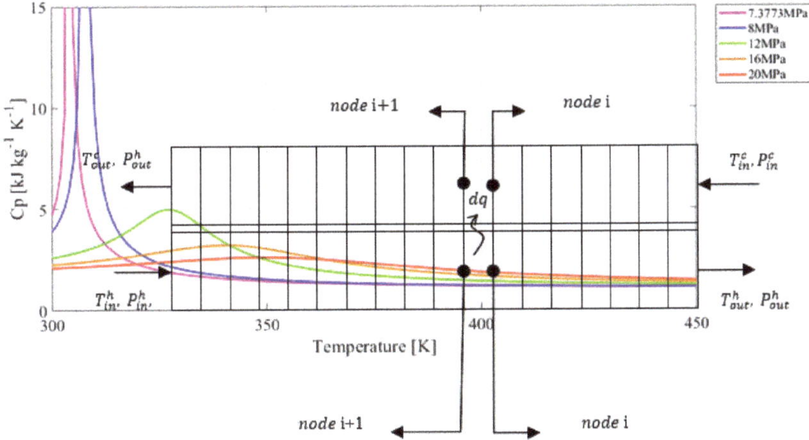

Figure 3. Discretized model of the heat exchanger to capture the effect of variation in the properties of sCO$_2$.

Macchi and Perdichizzi [14] used the turbine SP to determine the expander size, which for the sCO_2 re-compression Brayton cycle, is given by

$$SP_T = \frac{\sqrt{V_{8s}}}{\sqrt[4]{\Delta h_s}} = \frac{\sqrt{m_{CO_2} v_{8s}}}{\sqrt[4]{h_7 - h_{8s}}}, \tag{27}$$

where V_{8s} is the isentropic value of the volume flow rate of the working fluid at the turbine exit, and Δh_s is the isentropic specific enthalpy drop.

The main compressor and re-compression compressor SP values can be respectively defined as

$$SP_{MC} = \frac{\sqrt{m_{CO_2} x v_1}}{\sqrt[4]{h_{2s} - h_1}}, \tag{28}$$

$$SP_{RC} = \frac{\sqrt{m_{CO_2}(1-x)v_{10}}}{\sqrt[4]{h_{5s} - h_{10}}}. \tag{29}$$

For the system under study,

$$SP_{Total} = SP_T + SP_{MC} + SP_{RC}. \tag{30}$$

Hence,

$$(SP)_{Total} = \frac{\sqrt{m_{CO_2}v_{8s}}}{\sqrt[4]{h_7 - h_{8s}}} + \frac{\sqrt{m_{CO_2}xv_1}}{\sqrt[4]{h_{2s} - h_1}} + \frac{\sqrt{m_{CO_2}(1-x)v_{10}}}{\sqrt[4]{h_{5s} - h_{10}}}. \tag{31}$$

2.5. Properties of sCO$_2$

The selection of the working fluid has an important role in optimizing the performance of the system. Efficiency, cost and environmental aspects are key performance metrics that need to be considered before selecting a working fluid. Amongst the desired characteristics are an ozone depletion potential (ODP) of 0 and a low global warming potential (GWP) value. Table 2 lists the properties of CO_2, which is the working fluid used for the system in the study. Moreover, CO_2 is a non-toxic, cheap and easily available substance. A point of interest is that CO_2 has a critical temperature that is very close to the ambient temperature; thus, it is relatively easy for it to be used in the supercritical state, and it can be matched to a number of power cycles. The properties of CO_2 show large variations near their critical points, which is highly advantageous for the compression process. With a small increase in temperature, the energy of the fluid increases significantly during the compression process, thereby making it highly efficient [10].

Table 2. Typical properties of CO_2.

Molar Mass (g/mol)	Critical Pressure (MPa)	Critical Temperature (°C)	ODP	GWP
44.01	7.4	31.0	0	1

3. Results and Discussion

3.1. Thermal Efficiency (η_{th})

Figure 4 shows thermal efficiency, η_{th} of the sCO_2 system. As expected, the increment of T_7 increases the efficiency of the system, meaning that more power output can be derived for the same heat input. The optimal value for the split mass fraction (x) is the one for which the thermal efficiency shows the maximum value for each T_7 case. Moreover, as T_7 increases, the optimal value of x slightly shifts to the right. The best efficiency is at $x = 0.74$ and $T_7 = 700$ °C.

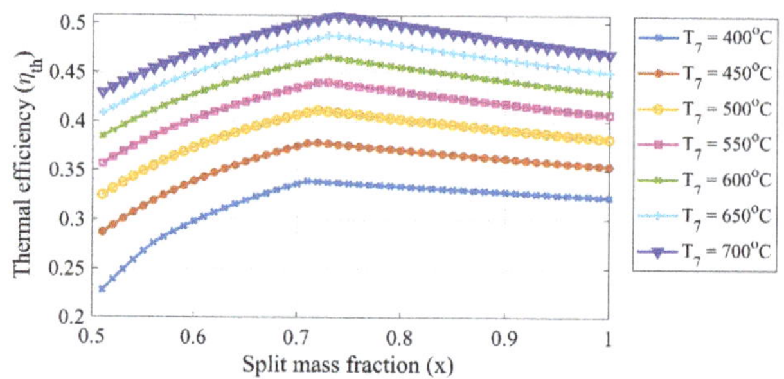

Figure 4. The thermal efficiency of the sCO_2 cycle at different turbine inlet temperatures (T_7).

3.2. Exergy Efficiency (η_{ex}) and Irreversibility Ratio (IR)

Figure 5 presents the *IR* and the η_{ex} values for the T_7 of 700 °C. The optimal value for *x* is the one for which η_{ex} shows the maximum value. The *IR* is the minimum, and at the same time, η_{ex} is the maximum at $x = 0.74$.

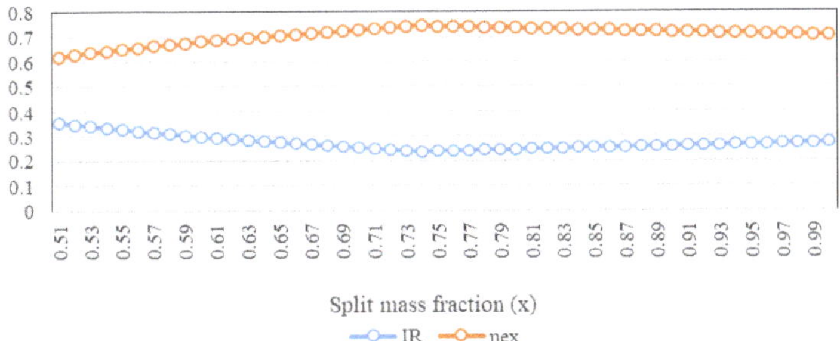

Figure 5. Exergy efficiency (η_{ex}) and irreversibility ratio (*IR*) of the sCO_2 cycle at a turbine inlet temperature (T_7) of 700 °C.

3.3. Exergy Destruction in Different Components

The total exergy destruction of the sCO_2 re-compression Brayton cycle increased with an increase in T_7, as shown in Figure 6. One of the reasons for this is the higher entropy generation at higher T_7 values causing comparatively higher exergy destruction in the turbine. Figure 7 presents the exergy destruction in different components of the sCO_2 cycle at $T_7 = 700$ °C and $x = 0.74$. It can be observed that the gas cooler shows the maximum exergy destruction because a lot of exergy is destroyed when the sCO_2 is cooled down from State 10 to State 1. Figure 8 shows that when the outlet pressure of the compressor is high, the exergy destruction is also comparatively high. Moreover, in the summer months, the exergy destruction was high because of high ambient and water temperatures. Figure 9 shows that as the compressor inlet temperature is kept to a lower value (near the critical point), the exergy destruction is minimized. The minimum exergy destruction of 390 kW was observed in January at a compressor inlet temperature of 35 °C and a compressor outlet pressure of 24 MPa.

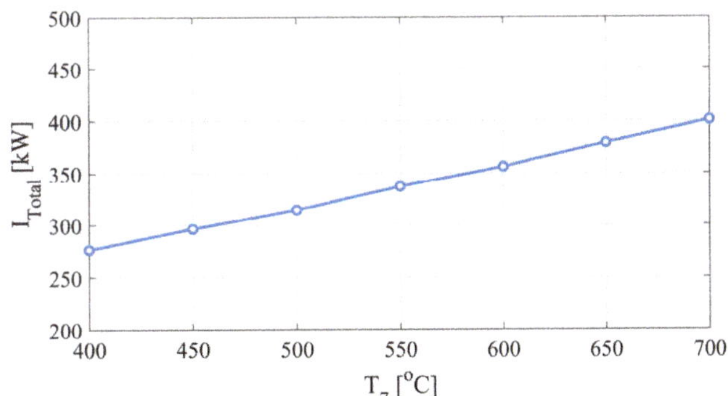

Figure 6. Total exergy destruction vs. turbine inlet temperature (T_7).

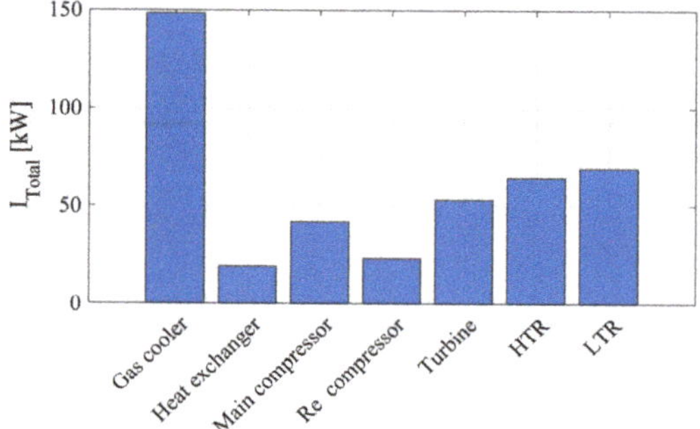

Figure 7. Exergy destruction in different components of the cycle at $T_7 = 700$ °C.

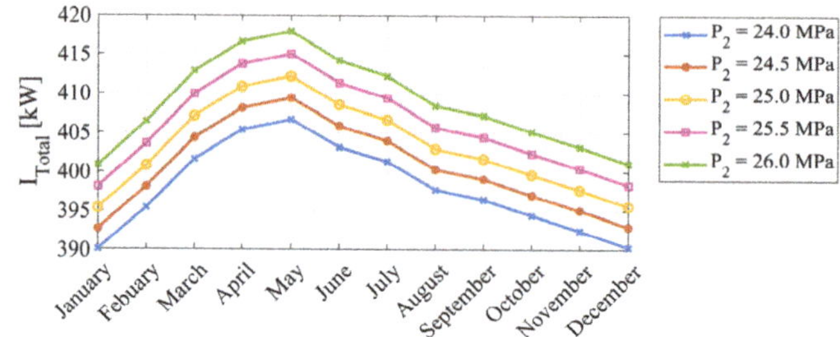

Figure 8. Total exergy destruction at different compressor outlet pressures (P_2).

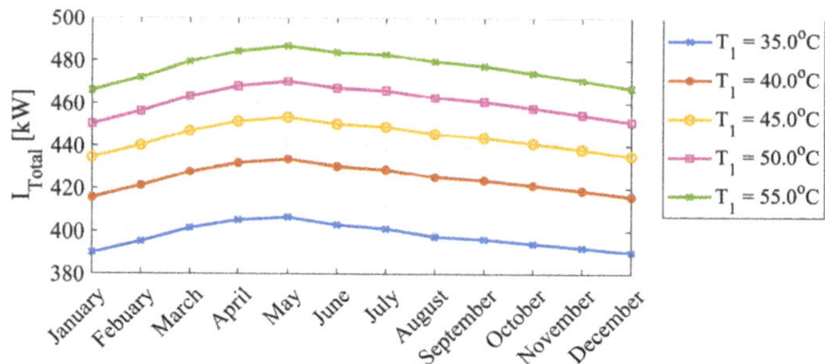

Figure 9. Total exergy destruction at different compressor inlet temperatures (T_1).

3.4. Heat Transfer Area Coefficient (UA) and Turbomachinery Size (SP) Values

UA and SP values are the parameters for the estimation of the cost of the system. UA value indicates the area required for heat exchangers, and thus indicates their cost. As the UA value increases, the total heat transfer area in the heat exchangers also increases, thereby increasing the investment and maintenance costs of the heat exchangers, and thus the system [21]. SP values indicate the size of the turbomachinery in the system; thus, a lower value is desirable.

Figure 10 illustrates the relationship of UA with the different heat exchangers in the sCO_2 re-compression Brayton cycle with different T_7 values. It should be noted that the UA value for the primary heat exchanger was the highest and increased at an increasing rate, whereas the UA value for the HTR showed a linear increase with respect to T_7. The net result is that for a high T_7 value, a large heat exchange area is required when considering all of the heat exchangers in the system. The gas cooler showed the lowest UA value and was not sensitive to T_7.

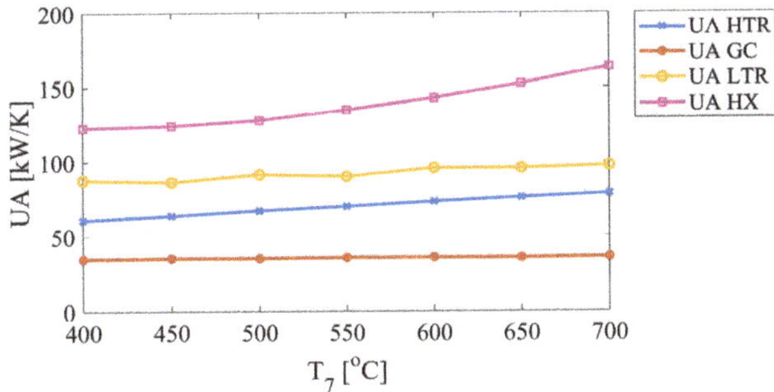

Figure 10. Heat transfer area coefficient (*UA*) vs. turbine inlet temperature (T_7).

Figure 11 shows the effect of T_1 on the heat exchanger sizes, as indicated by the *UA* value for each heat exchanger. It should be noted that the *UA* values of the primary heat exchanger, gas cooler and LTR all decrease with increasing T_1, whereas that of HTR shows only a slight increase. As shown in Figure 12, the net result was a decrease in the UA_{Total} value with increasing T_1, which implies that the total area required for heat exchange decreases, and smaller heat exchanger sizes will be required for higher compressor inlet temperatures (which are much higher than the critical temperature value).

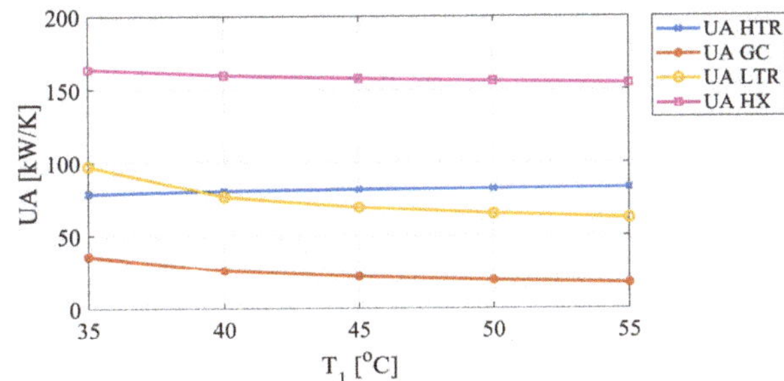

Figure 11. Heat transfer area coefficient (*UA*) vs. compressor inlet temperature (T_1).

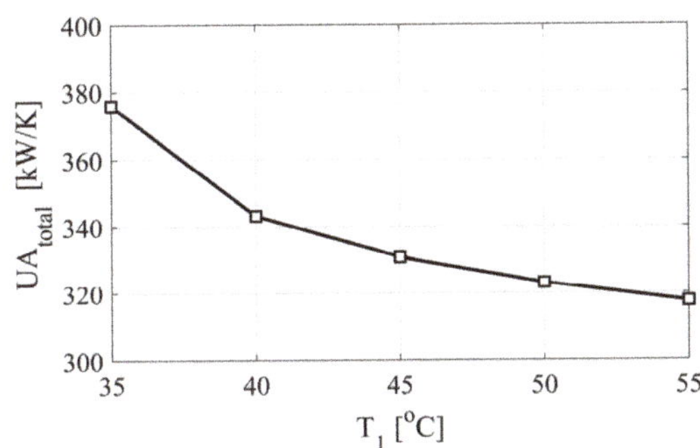

Figure 12. Total heat transfer area coefficient (UA_{Total}) vs. compressor inlet temperature (T_1).

Figure 13 shows the relationship between UA_{Total} and x in the sCO_2 re-compression Brayton cycle. The plot shows a maximum UA_{Total} value at $x = 0.74$, which is the same point at which η_{th} was maximum. This implies that the size, and thus, the cost of the heat exchangers in the system, will need to be high to achieve the maximum efficiency.

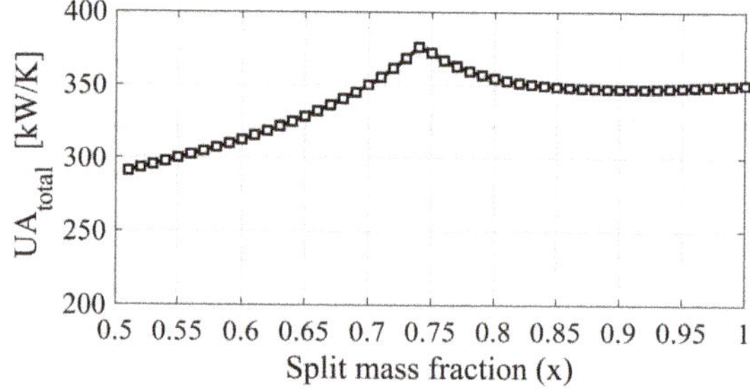

Figure 13. Total heat transfer area coefficient (UA_{Total}) vs. split mass fraction (x) at $T_7 = 700\ °C$.

Figures 14 and 15 show the SP values for the turbine, main compressor and re-compressor in the sCO_2 cycle. The SP of the turbine increases at higher T_7 values, implying that a large turbine should be used for high T_7 values. Moreover, the SP values of the main compressor and re-compressor are much less sensitive to T_7, meaning that the same compressor size can be used when T_7 is either low or high. The main compressor has a higher SP value than the re-compressor. Figure 15 shows the variation in SP of the different turbomachinery components with T_1. The SP of the turbine does not depend on T_1, as can be seen by the straight line. However, the SP values of the main compressor and re-compressor increase with an increase in T_1. Moreover, the increase in the SP value for the main compressor is more pronounced than that of the re-compressor. Hence, relatively large compressor sizes will be required at a high T_1.

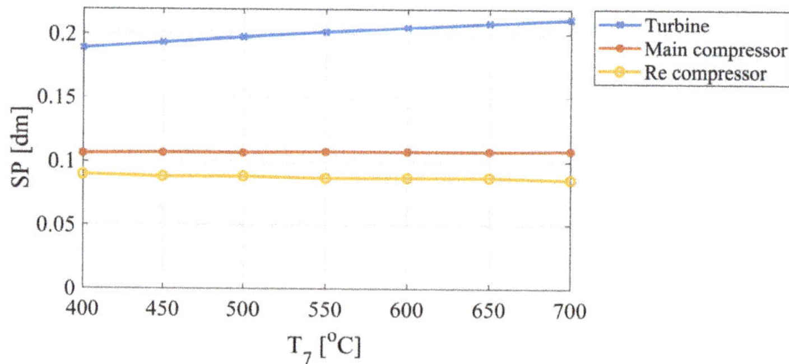

Figure 14. Size parameter (SP) vs. turbine inlet temperature (T_7).

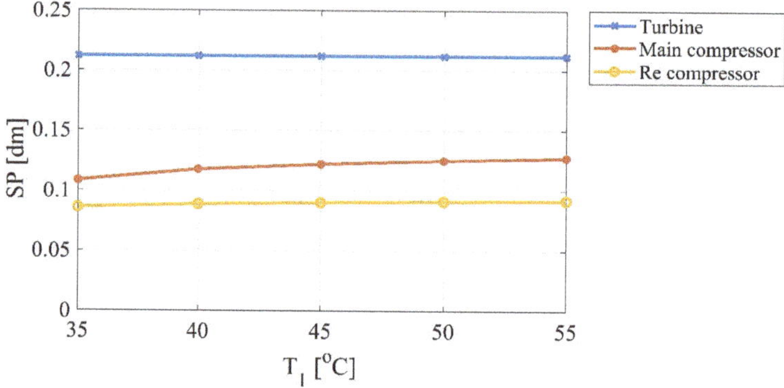

Figure 15. Size parameter (SP) vs. compressor inlet temperature (T_1).

4. Conclusions

The present study investigated seasonal performance analysis of sCO_2-based re-compression Brayton power system. From the analysis, the following conclusions were drawn:

- The thermal performance of the considered system is highly sensitive to the turbine's inlet temperate and split mass fraction. For every set of imposed boundary conditions, there exists an optimal value of split mass fraction that tends to minimize the temperature difference between LTR outlet and HTR inlet. It can be concluded that the optimal values of split mass fraction increase with the increase of turbine inlet temperature. Furthermore, the optimal value of the split mass fraction is found as 0.74, corresponding to the turbine inlet temperature of 700 °C for which cycle's efficiency (η_{th}) is 49%.
- The maximum of the cycle's efficiency comes at high total heat transfer area coefficient (UA_{Total}) values that in turn increase the cost of the heat exchanger. On the other hand, the cost of the heat exchanger can be reduced by moving the compressor's inlet temperature away from the critical temperature or by lowering the turbine inlet temperature. However, in both scenarios, the cycle's thermal efficiency shall be compromised. Therefore, an optimal trade-off can be established between UA_{Total} and η_{th} depending on the design objective.
- Exergy destruction is high for high turbine and main inlet temperatures and compressor inlet pressures. The gas cooler is the biggest source of exergy destruction in the cycle. Moreover, exergy destruction for the summer seasons was higher than the winter seasons. The system exergy destruction (I_{Total}) has the lowest value of 390 kW at 35 °C and 24 MPa in the month of January.
- The size parameter (SP) of the turbine increases with increasing turbine inlet temperature, whereas that of the compressor shows an increase when the main compressor inlet temperature increases. This implies that the total size parameter (SP_{Total}) value of the system increases with increasing turbine and compressor inlet temperatures, which the increase the cost of the turbomachinery under these conditions.

5. Future Work

- For the sCO_2 re-compression Brayton cycle, the thermal efficiency can be further increased if the heat available at the gas cooler is utilized to generate more power. This can be achieved by using the sCO_2 cycle in a cascade arrangement with an ORC as a bottoming cycle.
- A tri-generation system can be designed such that the multiple benefits of power, heating and cooling can be obtained from the system. Combined power and heating/cooling systems have the advantage of increasing efficiency by bringing multiple benefits into play.
- The performance of the system can be further improved if the system is optimized using a genetic algorithm such that the investment/maintenance costs of the system are minimized, and at the same time, the efficiency of the system is maximized. For this purpose, multi-objective optimization can be used using variables such as UA_{Total} and SP_{Total} to minimize the cost of the heat exchangers and turbomachinery in the system. Furthermore, η_{th} and η_{ex} can be used to maximize the efficiency of the system whilst minimizing the exergy destruction due to irreversibilities. The input variables within the most suitable bounded constraints can be defined in the optimization code.

Author Contributions: M.S.S. did the simulation analysis and drafted the manuscript. M.S. did heat exchanger analysis and edited the manuscript. M.-H.K. supervised the research and edited the manuscript. The authors read and approved the manuscript. All authors have read and agreed to the published version of the manuscript.

Funding: This research received no external funding.

Conflicts of Interest: The authors declare no conflict of interest.

Nomenclature

Symbols and abbreviations

Cp	Specific heat capacity
e	Specific exergy of flow [J/kg]
E	Exergy rate [W]
h	Specific enthalpy [J/kg]
I	Exergy destruction [W]
IR	Irreversibility ratio of the system
m	Mass flow rate [kg/s]
P	Pressure [Pa]
Q	Heat rate [W]
s	Specific entropy [J/kg.K]
SP	Turbomachinery size factor
SP	Size parameter [m]
T	Temperature [K]
UA	Heat transfer area coefficient [W/K]
W	Power [W]
x	Split mass fraction

Abbreviations

GC	Gas cooler
HTR	High temperature recuperator
HX	Primary heat exchanger
LTR	Low temperature recuperator
max	Maximum
min	Minimum
sCO_2	Supercritical Carbon dioxide
ΔT_M	Log mean temperature difference

Greek symbols

η	Efficiency
ε	Recuperator effectiveness
θ	Temperature difference

Subscripts

c	Cold side
CO_2	Carbon dioxide
h	Hot side
HTR	High temperature recuperator
HX	Primary heat exchanger
in	Property at inlet
LTR	Low temperature recuperator
MC	Main compressor
out	Property at outlet
ex	Exergy
RC	Re-compressor
s	Isentropic process
th	Thermal
T	Turbine
w	Water
0	Dead state
$1,2,3,$	State points

References

1. IPCC. *Climate Change 2014: Synthesis Report. Contribution of Working Groups I, II and III to the Fifth Assessment Report of the Intergovernmental Panel on Climate Change*; IPCC: Geneva, Switzerland, 2014; ISBN 9789291691432.
2. Mudasar, R.; Aziz, F.; Kim, M.-H. Thermodynamic analysis of organic Rankine cycle used for flue gases from biogas combustion. *Energy Convers. Manag.* **2017**, *153*, 627–640. [CrossRef]
3. Yağli, H.; Koç, Y.; Koç, A.; Görgülü, A.; Tandiroğlu, A. Parametric optimization and exergetic analysis comparison of subcritical and supercritical organic Rankine cycle (ORC) for biogas fuelled combined heat and power (CHP) engine exhaust gas waste heat. *Energy* **2016**, *111*, 923–932. [CrossRef]
4. Nasir, M.T.; Kim, K.C. Working fluids selection and parametric optimization of an Organic Rankine Cycle coupled Vapor Compression Cycle (ORC-VCC) for air conditioning using low grade heat. *Energy Build.* **2016**, *129*, 378–395. [CrossRef]
5. Liu, X.; Wei, M.; Yang, L.; Wang, X. Thermo-economic analysis and optimization selection of ORC system configurations for low temperature binary-cycle geothermal plant. *Appl. Therm. Eng.* **2017**, *125*, 153–164. [CrossRef]
6. Mudasar, R.; Kim, M.-H. Experimental study of power generation utilizing human excreta. *Energy Convers. Manag.* **2017**, *147*, 86–99. [CrossRef]
7. Singh, H.; Mishra, R.S. Performance evaluation of the supercritical organic rankine cycle (SORC) integrated with large scale solar parabolic trough collector (SPTC) system: An exergy energy analysis. *Environ. Prog. Sustain. Energy* **2018**, *37*, 891–899. [CrossRef]
8. Saeed, M.; Kim, M. Analysis of a recompression supercritical carbon dioxide power cycle with an integrated turbine design/optimization algorithm. *Energy* **2018**, *165*, 93–111. [CrossRef]
9. Ahn, Y.; Bae, S.J.; Kim, M.; Cho, S.K.; Baik, S.; Lee, J.I.; Cha, J.E. Review of supercritical CO_2 power cycle technology and current status of research and development. *Nucl. Eng. Technol.* **2015**, *47*, 647–661. [CrossRef]
10. Brun, K.; Friedman, P.; Dennis, R. (Eds.) *Fundamentals and Applications of Supercritical Carbon Dioxide (sCO$_2$) Based Power Cycles*; Woodhead Publishing, Elsevier Science: Cambridge, UK, 2017; ISBN 9780081008058.
11. Crespi, F.; Gavagnin, G.; Sánchez, D.; Martínez, G.S. Supercritical carbon dioxide cycles for power generation: A review. *Appl. Energy* **2017**, *195*, 152–183. [CrossRef]
12. Saeed, M.; Khatoon, S.; Kim, M.-H. Design optimization and performance analysis of a supercritical carbon dioxide recompression Brayton cycle based on the detailed models of the cycle components. *Energy Convers. Manag.* **2019**, *196*, 242–260. [CrossRef]
13. Aziz, F.; Mudasar, R.; Kim, M.-H. Exergetic and heat load optimization of high temperature organic Rankine cycle. *Energy Convers. Manag.* **2018**, *171*, 48–58. [CrossRef]
14. Macchi, E.; Perdichizzi, A. Efficiency prediction for axial-flow turbines operating with non conventional fluids. *J. Eng. Power* **1981**, *103*, 718–724. [CrossRef]
15. Aziz, F.; Salim, M.S.; Kim, M.-H. Performance analysis of high temperature cascade organic Rankine cycle coupled with water heating system. *Energy* **2019**, *170*, 954–966. [CrossRef]
16. Patel, B.; Desai, N.B.; Kachhwaha, S.S. Optimization of waste heat based organic Rankine cycle powered cascaded vapor compression-absorption refrigeration system. *Energy Convers. Manag.* **2017**, *154*, 576–590. [CrossRef]
17. MATLAB—MathWorks-R2017a (Version 9.2). 2017. Available online: https://www.mathworks.com/products/new_products/release2017a.html (accessed on 12 February 2020).
18. Lemmon, E.; Mc Linden, M.; Huber, M. NIST Reference Fluid Thermodynamic and Transport Properties Database: REFPROP Version 9.1, NIST Standard Reference Database 23. 2013. Available online: http://www.boulder.nist.gov (accessed on 25 December 2017).
19. Busan Ambient Temperature Data. Available online: https://en.climate-data.org/asia/south-korea/busan/busan-4114/ (accessed on 15 September 2019).
20. Moran, M.J.; Shapiro, H.N.; Boettner, D.D.; Bailey, M.B. *Principles of Engineering Thermodynamics*, 7th ed.; John Wiley & Sons: Hoboken, NJ, USA, 2012.
21. Gao, H.; Liu, C.; He, C.; Xu, X.; Wu, S.; Li, Y. Performance analysis and working fluid selection of a supercritical organic Rankine cycle for low grade waste heat recovery. *Energies* **2012**, *5*, 3233–3247. [CrossRef]

22. Saeed, M.; Kim, M.-H. Thermal and hydraulic performance of SCO$_2$ PCHE with different fin configurations. *Appl. Therm. Eng.* **2017**, *127*, 975–985. [CrossRef]

23. Dostal, V.; Driscoll, M.J.; Hejzlar, P. *A Supercritical Carbon Dioxide Cycle for Next Generation Nuclear Reactors, MIT-ANP-TR-100, Advanced Nuclear Power Technology Program Report*; Massachusetts Institute of Technology: Cambridge, MA, USA, 2004.

Article

A Supercritical CO_2 Waste Heat Recovery System Design for a Diesel Generator for Nuclear Power Plant Application

Jin Ki Ham [1,2], Min Seok Kim [2], Bong Seong Oh [1], Seongmin Son [1], Jekyoung Lee [2] and Jeong Ik Lee [1,*]

[1] Department of Nuclear and Quantum Engineering, Korea Advanced Institute of Science and Technology, 373-1 Guseong-dong Yuseong-gu, Daejeon 34141, Korea; jkham@kaist.ac.kr (J.K.H.); bongseongoh@kaist.ac.kr (B.S.O.); ssm9725@kaist.ac.kr (S.S.)
[2] Korea Shipbuilding & Offshore Engineering, 75 Yulgok-ro, Jongno-gu, Seoul 03058, Korea; minseok.kim@ksoe.co.kr (M.S.K.); jekyoung.lee@ksoe.co.kr (J.L.)
* Correspondence: jeongiklee@kaist.ac.kr; Tel.: +82-42-350-3829; Fax: +82-42-350-3810

Received: 13 November 2019; Accepted: 5 December 2019; Published: 9 December 2019

Abstract: After the Fukushima accident, the importance of an emergency power supply for a nuclear power plant has been emphasized more. In order to maximize the performance of the existing emergency power source in operating nuclear power plants, adding a waste heat recovery system for the emergency power source is suggested for the first time in this study. In order to explore the possibility of the idea, a comparison of six supercritical carbon dioxide (S-CO_2) power cycle layouts recovering waste heat from a 7.2 MW alternate alternating current diesel generator (AAC DG) is first presented. The diesel engine can supply two heat sources to the waste heat recovery system: one from exhaust gas and the other from scavenged air. Moreover, a sensitivity study of the cycles for different design parameters is performed, and the thermodynamic performances of the various cycles were evaluated. The main components, including turbomachinery and heat exchangers, are designed with in-house codes which have been validated with experiment data. Based on the designed cycle and components, the bottoming S-CO_2 cycle performance under part load operating condition of AAC DG is analyzed by using a quasi-steady state cycle analysis method. It was found that a partial heating cycle has relatively higher net produced work while enjoying the benefit of a simple layout and smaller number of components. This study also revealed that further waste heat can be recovered by adjusting the flow split merging point of the partial heating cycle.

Keywords: emergency diesel generator; supercritical carbon dioxide cycle; waste heat recovery system; bottoming cycle

---

1. Introduction

The earthquake that occurred in Japan on 11 March 2011 induced a tsunami with several waves whose height reached more than ten meters. Unfortunately, the Fukushima Daiichi nuclear power plant could not maintain its integrity after the large earthquake and tsunami. One of the major reasons was that the electric power supply lines to the site as well as the operational and safety infrastructure on the site were severely damaged. As a result, the on-site and off-site electrical power loss led to the loss of the cooling function in three reactor units and the spent fuel storage pools. Despite of the follow-up efforts at the Fukushima Daiichi nuclear power plant staff, the reactor buildings in Units 1, 3 and 4 were breached due to a hydrogen explosion [1].

After the Fukushima accident, various countries operating nuclear power plants have revised their policies while adding more safety systems. In Korea, 50 short and long-term action items were identified and implemented to respond to the post-Fukushima accident nuclear safety concerns. Among these items, access to a vehicle with generators and batteries is a key lesson from the long term complete Station Black Out (SBO) accident that led to the core melting of the Fukushima nuclear power plant [1]. Table 1 shows some of the improvements implemented after the Fukushima accident.

Table 1. Safety improvements after the Fukushima accident (recreation of the figure in Ref. [1]).

	Current Status	Improvements
Electric Power System	* 2 EDGs/unit - Loss of offsite power * 1 AAC DG/2 or 4 units - Loss of cooling function (SBO)	* Movable vehicle for generator and batteries - ~2014, All NPPs
Cooling System	* Redundancy (2 trains) * SFP has multiple sources	* Prepare supplementary methods - fire truck, etc. - ~2013, All NPPs
Fire Protection System	* Fire hazards analysis/10 years * Fire protection plans	* Improving the firefighting plan * Improving fire protection facility - ~2015, All NPPs

Even though these safety actions are new suggestions adopted in operating nuclear power plants, improving nuclear safety has always been emphasized, even before Fukushima accident. Hence, in this study, an additional power supply system without substantially revising the layout and systems of an existing nuclear power plant will be proposed.

Currently, Korean nuclear power plants have two emergency diesel generators (EDG) and a non-class 1E diesel generator for each unit. Furthermore, an alternate alternating current diesel generator (AAC DG) for the emergency power supply that can be used for multiple on site units is installed. AAC DG is a standby system to supply the emergency power to the nuclear power plant when both EDGs are unavailable during accident conditions. However, AAC DG only provides emergency electrical power to the class 1E safety system due to its limited rated power and fuel tank size. Thus, if the AAC DG can generate more power without revising current layout of a nuclear system, the additional power can be used for further improving the nuclear safety. The proposed idea to fulfill this mission is adopting a waste heat recovery system to generate electricity further. The waste heat is generated from AAC DG. Figure 1 shows the configuration of the schematic emergency power supply system of the nuclear power plant.

The target nuclear power plant to apply a waste heat recovery system is APR 1400 that is the abbreviation of "Advanced Power Reactor with an electrical power output of 1400 MW". It is a pressurized water-cooled reactor developed in Korea. The main features of the APR 1400 are enhanced plant safety, economically favorable and convenient operation and maintenance. In terms of the electrical system of APR 1400, the electrical buses are separated into Class 1E A and Class 1E B. Class 1E is the systems that provides essential electric power to reactor shutdown, containment isolation, emergency core cooling, removal of residual heat, and preventing serious leakage of radioactive material to the environment. The list of Class 1E electrical loads is given in Table 2 [2].

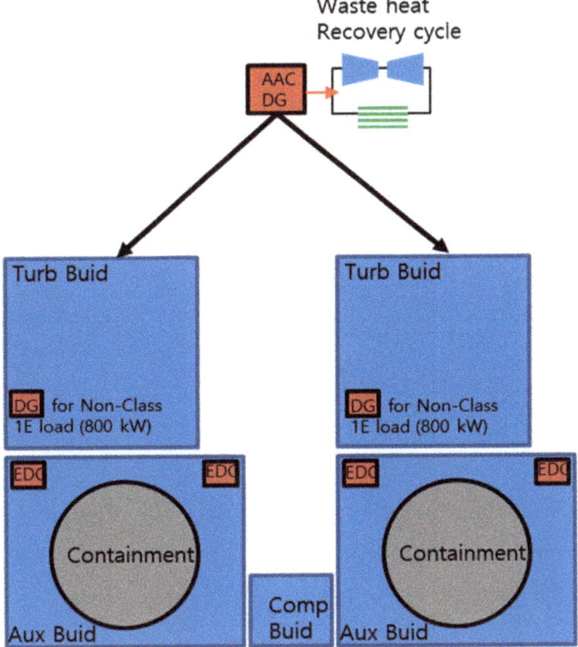

Figure 1. Schematic diagram of nuclear power plant and emergency power supply systems.

Table 2. Class 1E load and sequence at LOOP condition.

Class 1E Components	Loading Sequence at LOOP	Power (kW)
Load Sequence Group ×2	0.0 s	694.9 × 2
Safety injection pump 1	5.0 s	715
Safety injection pump 3	10.0 s	715
Motor driven AFW pump	15.0 s (if required)	930.9
Containment spray pump (That interlocks with shutdown cooling pump)	20.0 s	738.1
Component cooling water pump	25.0 s	1225
Essential service water pump	30.0 s	636.8
Essential central chiller	35.0 s	806.0
Essential ESW intake structure/ CCW heat exchanger building chiller	40.0 s	304.9
Total Diesel Load on DBA/LOOP excluding Manual Load		7461.7

In case of failure of EDG, AAC DG supplies power to the Class 1E load instead of EDG.

For redundancy, independent EDGs supply power to each bus in LOOP and SBO accidents. Another 800 kW diesel generator in the turbine building is installed for supplying power to non-Class 1E as well if the EDG system becomes unavailable, AAC DG is electrically connected to operate safety systems instead of failed EDG. Table 3 contains the actual diesel generator specifications and vendors of APR 1400 which is constructed in Korea. In this study, AAC diesel generators of Shin Hanul is selected as the reference system. Figure 2 and Table 4 represent the schematic diagram and specification of diesel generator, respectively.

Table 3. Status of EDG & AAC DG of APR 1400 [3].

NPP site	Unit	No. of DG	Power (kW)	Vendor (Engine/Generator)
Shin Kori	EDG 3/4	4	8000	Doosan-MDT: 16PC2.6B, Alstom
	AAC DG 3, 4	1	7200	Doosan-MDT: 16PC2.6B, Alstom
Shin Hanul	EDG 1/2	4	7200	Doosan-SEMT: 16PC2-5V400, Alstom
	AAC DG 1, 2	1	7200	Doosan-MDT: 18V32/40, Hyundai

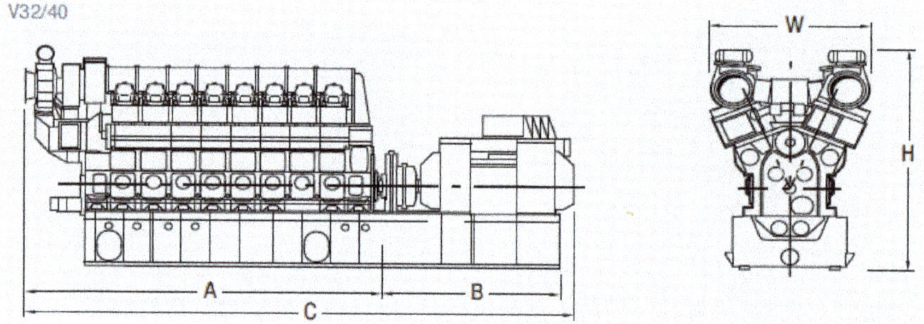

Figure 2. Schematic diagram of the Doosan-MDT: 18V32/40 for AAC diesel generator.

Table 4. Specifications of the Doosan-MDT: 18V32/40 for AAC diesel generator.

Specification	
Engine	18V/32/40
A (mm)	8300
B (mm)	4450
C (mm)	12,750
H (mm)	5240
W (mm)	3500
Weight (t)	139
Exhaust gas data	
Temperature at turbine outlet	306 °C
Mass flowrate	15.1 kg/h
Volume flowrate	92,700 m^3/h
Pressure (abs.)	1.03 bar
Permissible pressure drop after turbine	<0.03 bar
Scavenged air data	
Temperature at the compressor inlet	25 °C
Temperature at the air cooler outlet	42 °C
Mass flowrate	14.7 kg/h
Volume flowrate	48,000 m^3/h
Pressure (abs.)	3.2 bar

From Table 4, it can be observed that the thermal potential of waste heat is quite high because a turbine outlet temperature is 306 °C, and the mass flowrate is 15.1 kg/h which is equivalent to about 3 MWth heat so that it is quite valuable to use this high potential waste heat as an additional power for operating a safety system. For example, additional electric power from the waste heat recovery system can be used for extended operation of AAC DG while using the same amount of fuel or installing more safety components for improving redundancy or diversity to utilize additional power.

Consequently, the additional power from the AAC DG waste heat recovery system can be utilized in the nuclear safety system without revising the existing nuclear power plant's layout. In this paper, the supercritical carbon dioxide (S-CO$_2$) bottoming cycles are studied in respect of best applicable options [4]. According to the review of the previous studies on the S-CO$_2$ cycles, it was verified that the S-CO$_2$ cycle has a strong potential to outperform the conventional steam cycle or organic Rankine cycle, particularly in waste heat recovery applications [5]. The major reasons are as follows: high efficiency at a moderate turbine inlet temperature (450–750 °C) which significantly decreases maintenance and material problems, compact turbomachinery and heat exchangers which save the initial investment, and a simple layout which significantly reduces the overall footprint of the power plant [6]. The main feature of the S-CO$_2$ Brayton cycle is lower compression work than the other gas Brayton cycles because of S-CO$_2$ fluid's its high density and low compressibility near the critical point. Therefore, this characteristic makes the S-CO$_2$ cycle having fewer material issues than water and consequently leads to a higher turbine inlet temperature [7]. Moreover, the cooling and chemistry control system of S-CO$_2$ cycle are relatively simple than the steam Rankine cycle, the operation cost and the whole footprint of the power plant can be greatly decreased [8]. Because of these advantages, the S-CO$_2$ cycle is recently considered as the potential power cycle of conventional and renewable energy systems such as fossil fuel power plants, concentrated solar power systems, geothermal power plants, fuel cells, next generation nuclear power plants, and ship propulsion application.

In this paper, a comparison of six S-CO$_2$ power cycle layouts recovering waste heat of s 7.2 MW AAC DG is presented in order to evaluate which S-CO$_2$ cycle layout is the most qualified for a diesel engine application. Moreover, a sensitivity study of the cycles with design parameters was carried out, and thermodynamic performance results of the cycles were evaluated. After that, key components of the waste heat recovery system will be designed, and off-design analysis of the designed S-CO$_2$ bottoming cycle will be presented.

2. S-CO$_2$ Bottoming Cycle Study for Diesel Generator

2.1. Cycle Layout and Analysis Method

It is well known that the turbine inlet temperature and pressure ratio have great effect on the efficiency of gas Brayton cycle. However, overall cycle efficiency and a heat recovery factor should be carefully adopted for the application of waste heat recovery system because the factors are directly combined to the work recovered from the amount of transferred waste heat to the bottoming cycles.

Martelli et al. [9] noted that to obtain the net efficiency of the heat recovery cycle (calculated by dividing net produced work by the total heat recovered through ideally cooling flue gases to ambient temperature), the product of the cycle efficiency (net produced work divided by recovered heat) and the heat recovery factor (recovered heat divieded by ideally recoverable heat) should be calculated. Hence, when optimizing heat recovery cycles, instead of maximizing the cycle efficiency, maximizing the product of cycle efficiency and heat recovery factor becomes more valid.

According to Heo et al. [10], the performance indicator of bottoming cycles can be simply expressed by adopting the waste heat recovery index (WHRI):

$$\mathrm{WHRI} = \chi \cdot \eta_{cycle} \tag{1}$$

$$\chi = \frac{\text{thermal power absorbed by the bottoming cycle}}{\text{total net thermal power recoverable dueto thermal power}}$$
$$= \frac{\dot{m}_{exhaust}\left(h\left(T_{exhaust,in},P_{exhaust,in}\right)-h\left(T_{exhaust,out},P_{exhaust,out}\right)\right)}{\dot{m}_{exhaust}\left(h\left(T_{exhaust,max},P_{exhaust,max}\right)-h(T_{ambient},P_{ambient})\right)} \tag{2}$$

$$\eta_{cycle} = \frac{\text{net work generated by the bottoming cycle}}{\text{thermal power absorbed by the bottoming cycle}}$$
$$= \frac{W_{turbine}-W_{compressor}}{\dot{m}_{exhaust}\left(h\left(T_{exhaust,in},P_{exhaust,in}\right)-h\left(T_{exhaust,out},P_{exhaust,out}\right)\right)} \tag{3}$$

$$\therefore \mathrm{WHRI} = \frac{\text{net work generated by the bottoming cycle}}{\text{total net thermal power recoverable due to thermal power}}$$
$$= \frac{W_{turbine}-W_{compressor}}{\dot{m}_{exhaust}\left(h\left(T_{exhaust,max},P_{exhaust,max}\right)-h(T_{ambient},P_{ambient})\right)} \tag{4}$$

where h(T, P) denotes the calculated enthalpy values from the given temperature and pressure values.

By utilizing this performance index, various bottoming cycle designs can be assessed with a more applicable framework especially for the waste heat recovery systems' performance. This indicator matches to the concept of cycle net efficiency, except that it evaluates the performance replacing the cycle heat input with maximum obtainable heat from the heat source for the bottoming cycle.

The WHRI values for various S-CO_2 cycle layouts are obtained. Parametric sensitivity of the design variables such as turbine outlet pressure (or pressure ratio), CO_2 mass flow rate, first compressor outlet pressure, and flow splits, with respect to cycle performance, is summarized. Furthermore, the cycle minimum pressure is placed as a major design parameter to be optimized.

An in-house code is used for the cycle optimization. The code is developed by the research team at KAIST, and it is named as KAIST-Closed Cycle Design (KAIST-CCD) code [11–13]. The code is based on MATLAB. Enthalpy and fluid properties for calculation in the code are referred from the NIST reference fluid thermodynamic and transport properties database (REFPROP) [14]. Figures 3 and 4 show the algorithm of the code structure and the partial heating cycle layout. This layout is chosen to illustrate the process of the cycle analysis. Firstly, the recuperator cold side inlet flow condition (point 5) and the heater 2 hot side inlet flow condition (point 10) are assumed. Through the component models (compressor, turbine, recuperator, heater, cooler, and mixing tee), the component inlet and outlet conditions are calculated. Then, the heat input from exhaust gas with exhaust inlet and outlet conditions is calculated. If the cycle calculation error remains above 10^{-5}, the assumed values are updated to the newly calculated conditions, and the cycle calculation is repeated. If the error is less than 10^{-5}, the code calculation terminates and prints output results. The partial heating cycle and other layouts are evaluated by the KAIST-CCD code through a similar algorithm.

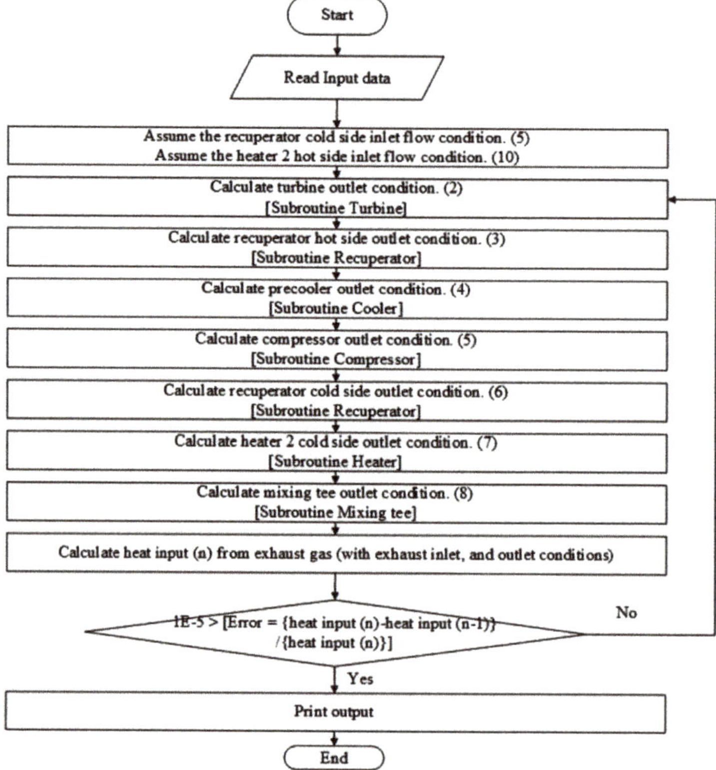

Figure 3. The cycle design code algorithm for the partial heating cycle.

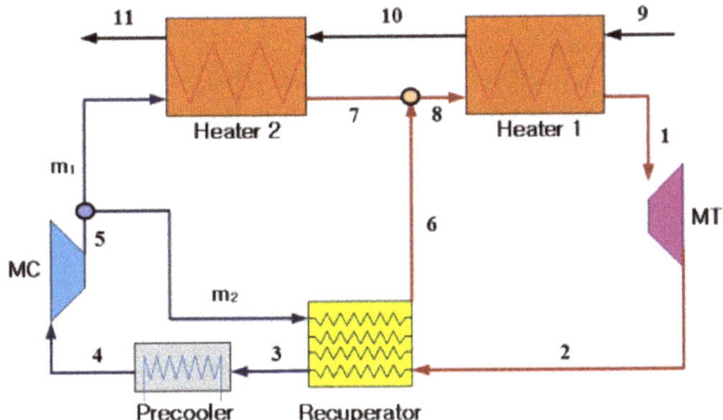

Figure 4. Layout of partial heating cycle for the algorithm illustration purpose.

The full power of the reference four strokes diesel engine is 9 MW with 720 or 750 rpm, but around 80% of full power is used for the emergency electrical load due to safety margin. Table 5 shows the main specification of the AAC DG which influences the design of the S-CO$_2$ bottoming cycle. Table 6 displays the values of exhaust gas source, properties, and mole fraction of each species [15]. The modeling of each S-CO$_2$ cycle adopts the exhaust gas properties of the topping cycle to evaluate the power output of the S-CO$_2$ cycle.

Table 5. Specifications of AAC DG.

Exhaust Gas	
Power capacity	7.2 MWe (80% load)
Exhaust temperature at turbine outlet	306 °C
Pressure (abs.)	1.03 bar
Mass flow rate	15.1 kg/s
Scavenge Air	
Air temperature at compressor inlet	25 °C
Air temperature at compressor outlet	42 °C
Pressure (abs.)	3.2 bar
Mass flow rate	14.7 kg/s

Table 6. Exhaust gas properties, source, and mole fraction of each species.

	Species	Mole Fraction	Source
Major species	Nitrogen (N$_2$)	0.78	Inlet air
	Oxygen (O$_2$)	0.13	Inlet air
	Carbon dioxide (CO$_2$)	0.06	Oxidation of fuel carbon
	Water vapor (H$_2$O)	0.03	Oxidation of fuel hydrogen

2.2. Assumptions and Constraints

To observe the performance of various cycle layouts under the same exhaust gas condition, several constraints are assumed. Here, the maximum operating pressure is limited to 22.0 MPa. This value corresponds to the maximum pressure of the existing ultra-supercritical steam Rankine cycle which already operates under significantly high pressure, and its maximum temperature also surpasses 600 °C [16]. The maximum operating pressure is limited to 22.0 MPa to avoid high capital costs for the piping systems and to generate a realistic S-CO$_2$ power cycle model. The maximum temperature (turbine inlet temperature) falls below the exhaust temperature of the topping cycle. Finally, the minimum cycle temperature (compressor inlet temperature) is restricted to 25.0 °C so that the heat sink temperature does not fall too far from the general room temperature while the cycle fluid remains in the supercritical state [17].

The cycle variables are summarized in Table 7, including the cycle minimum pressure, pressure drop of the heat exchangers, turbomachinery efficiency and heat exchanger effectiveness. Particularly, turbomachinery efficiency and heat exchanger effectiveness were based on the reference in order to conduct sensitivity analyses of cycles. The previous research results have shown that the cost of heat exchanger sharply increases when the heat exchanger effectiveness value is above 0.95 [18]. The pressure drop values in CO_2 side of all heat exchangers were assumed to be 0.5% of the system pressure.

Table 7. Cycle variables of S-CO_2 bottoming cycles.

Content		Unit	Value
Heat source		-	Exhaust gas only or exhaust gas & scavenge air
Cycle minimum pressure			22.0
Pressure drop	Exhaust gas/Scavenge air	MPa	0.202 (=2bar)
	Others		0.101 (=1bar)
Temperature conditions	Air	°C	25
	Cooling water		25
	Min. temp. difference		10
	Compressor inlet		35
Turbomachinery efficiency	Turbine	%	88
	Compressor		75
Heat exchanger effectiveness	Exhaust gas	%	90
	Scavenge air		90
	Precooler		90
	Recuperator		95

When designing heat exchangers, in particular when calculating the heat transfer, it is known that the specific heat capacity property of the working fluid is very important because the specific heat capacity has a great influence on the heat exchanger effectiveness as well as cycle efficiency. Generally, the specific heat capacity is considered as a constant value due to a little change with pressure and temperature. However, this assumption is inappropriate for the supercritical fluid since the property of the specific heat capacity changes substantially near the critical point. In this study, all properties as well as the specific heat capacity were directly called from the REFPROP.

2.3. Sensitivity Analysis of S-CO_2 Cycles

In the S-CO_2 cycle research community, a study for the S-CO_2 bottoming cycle is mostly performed for the patent application; therefore, the patented cycle's optimization results are proprietary information which is not openly accessible to the researchers in this field. Thus, in this paper, the academic contributions are presenting the analysis results of various S-CO_2 cycle layouts for bottoming cycle application to generate database for the future study. For the waste heat recovery process, all the selected cycle layouts are free from patent and searched from open literatures as well as a newly suggested cycle by the authors. For this purpose, cycle layouts suggested by various S-CO_2 power system companies are excluded.

2.3.1. Simple Recuperation Cycle

Firstly, a simple recuperated cycle [19] is analyzed as a basic cycle to compare the cycle performance of other S-CO_2 cycles. Five different cycles were selected as candidates and number of components and design variables are summarized in Table 8. The simple recuperated cycle has 515 kWe net produced work (see Figure 5) which is the lowest among six cycles considered in this paper because EG heat exchanger only absorbs 2.55 MWth of waste heat due to the low temperature difference between hot side and cold side of EG heat exchanger. Although the simple recuperated cycle produces the lowest net work, its footprint is relatively small because of the lowest total exchanged heat in heat exchangers and the fewest number of components.

Table 8. The design variables and the number of components of S-CO$_2$ cycles.

Cycle Layout	The Number of HX/TB/CP	Design Variables
(1-1) Simple recuperation	3/1/1	CO$_2$ Mass flow rate, turbine outlet pressure
(1-2) Simple recuperation with SA heat	4/1/1	CO$_2$ Mass flow rate, turbine outlet pressure
(2-1) Recompression cycle with SA heat	5/1/2	CO$_2$ Mass flow rate, turbine outlet pressure, split ratio 1, split ratio 2
(3-1) Partial heating cycle	4/1/1	CO$_2$ Mass flow rate, turbine outlet pressure, split ratio
(3-2) Partial heating cycle with SA heat	4/1/1	CO$_2$ Mass flow rate, turbine outlet pressure, split ratio
(3-3) Modified partial heating cycle with SA heat	4/1/1	CO$_2$ Mass flow rate, turbine outlet pressure, split ratio

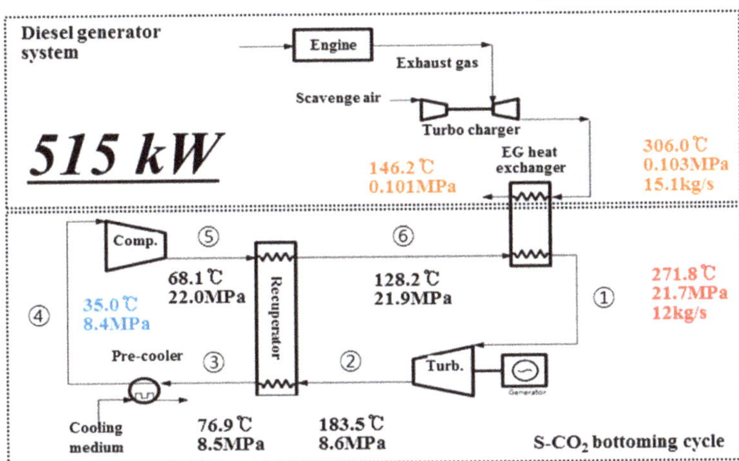

Figure 5. Configuration of simple recuperation cycle.

2.3.2. Simple Recuperation Cycle with SA Heat

The S-CO$_2$ power conversion system has considered the only single waste heat source from various power plants. However, diesel engine application has dual heat source from exhaust gas and scavenge air as shown in Figure 6. To additionally use the SA heat from AAC DG turbocharger, simple recuperation cycle was modified. It has one more waste heat exchanger than the simple recuperation cycle for SA heat absorption.

The T-s diagram of the simple recuperation cycle with SA heat is shown in Figure 7. Figure 8 shows the variation of net produced work with respect to the selected design variables. The simple recuperated cycle with SA heat has 12.8% WHRI and 568 kWe net produced work which is 1.2% and 53 kWe (+10.3%) higher than the simple recuperated cycle. By applying dual heat sources from exhaust gas and scavenge air, waste heat recovery is sharply increased from 2.55 MWth to 3.21 MWth (+25.9%) with one more heat exchanger.

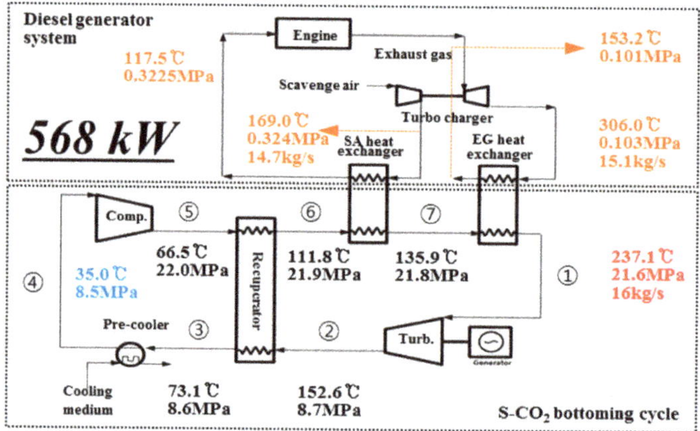

Figure 6. Configuration of simple recuperation cycle with SA heat.

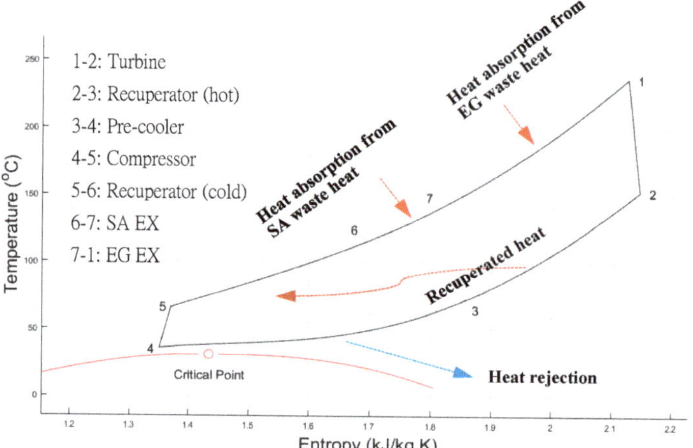

Figure 7. T-s diagram of simple recuperation cycle with SA heat.

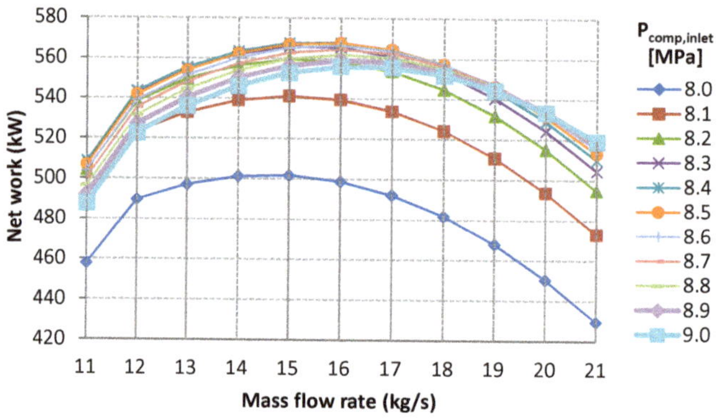

Figure 8. Net work dependence with CO_2 mass flow rate, and turbine outlet pressure.

2.3.3. Modified Recompression Cycle with SA Heat

According to the previous works [20,21], recompression cycle which has split flow and two different compressors is one of the representative layout of S-CO$_2$ cycle since it provides the highest efficiency. The split flow can minimize waste of available heat (see Figure 9) and it enhances overall cycle efficiency. However, verification for the suitability of a recompression cycle as a bottoming cycle is needed since the net produced work is a more important factor than the thermal efficiency in the design of bottoming cycle.

The T-s diagram of recompression cycle with SA heat is shown in Figure 10. The modified recompression cycle with SA heat has one more heat exchanger and one more compressor than the simple recuperation with SA heat. The net produced work variation with respect to the selected design variables is shown in the Figure 11. In this paper, flow split is expressed as $m_1 / (m_1 + m_2)$. As shown in Figure 11, the thermal efficiency is reduced but the heat input has the reverse trend as flow split decreases. In other words, higher net produced work is expected as flow split is closer to zero since the increase of heat input is greater than the decrease of thermal efficiency. If the flow split is equal to zero, this cycle is similar to a partial heating cycle with SA heat; therefore, the recompression cycle is not suitable for the bottoming cycle application.

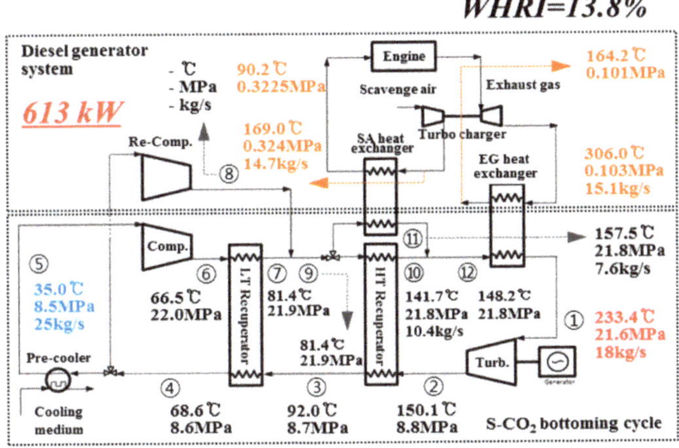

Figure 9. Configuration of recompression cycle with SA heat.

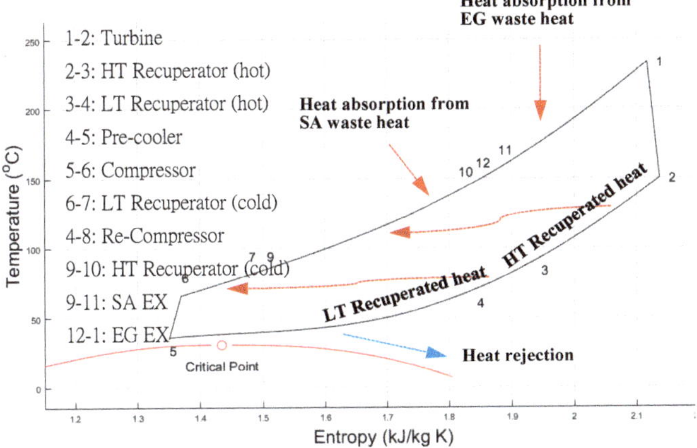

Figure 10. T-s diagram of recompression cycle with SA heat.

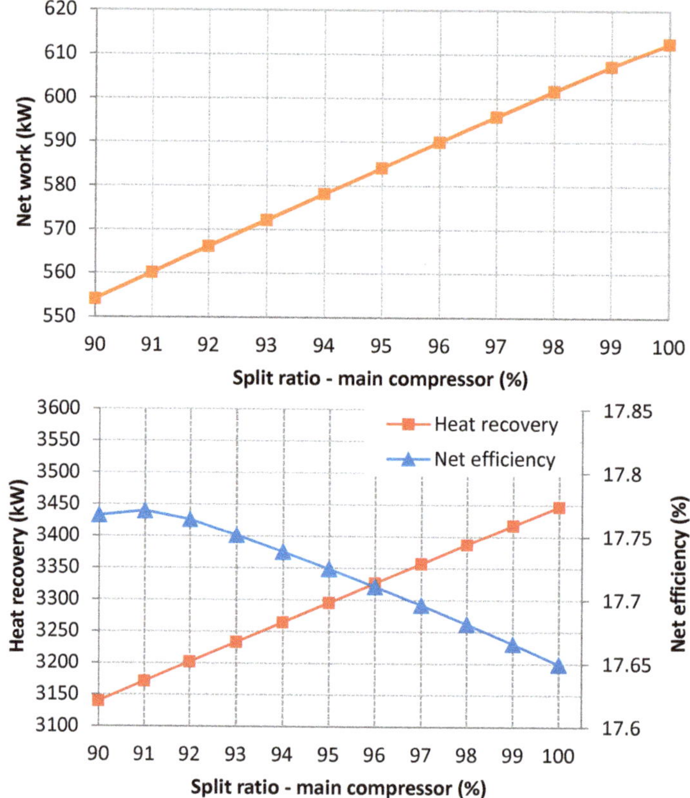

Figure 11. Net produced work dependency on flow split (upper), heat recovery and net efficiency dependency on flow split (bottom).

2.3.4. Partial Heating Cycle

The partial heating cycle [22] which has the same number of heat exchangers with the simple recuperation with SA heat is analyzed. Since it has a different design variable and flow split ratio from the simple recuperation with SA heat, the effect of flow split ratio has to be investigated again (see Figure 12). Figure 13 shows the T-s diagram of the partial heating cycle for the S-CO$_2$ bottoming cycle. The net produced work variation with respect to the selected design variables is shown in Figure 14.

Unlike the simple recuperation with SA heat, this cycle achieves quite high 607 kWe net produced work for its small number of components (only four heat exchangers). From the sensitivity analysis, the net produced work is maximized, and the optimal absorbed heat from the exhaust and recuperated heat is found by controlling the flow fraction to the recuperator and the second heater. As shown in Figure 12, a final exhaust gas temperature is 77.4 °C which is relatively lower temperature, and this makes higher absorbed heat from the exhaust gas of a diesel engine. The partial heating cycle is expected to show the best performance among the selected cycle layouts, because the partial heating cycle has a relatively large recuperated heat with a simple layout and a small number of components.

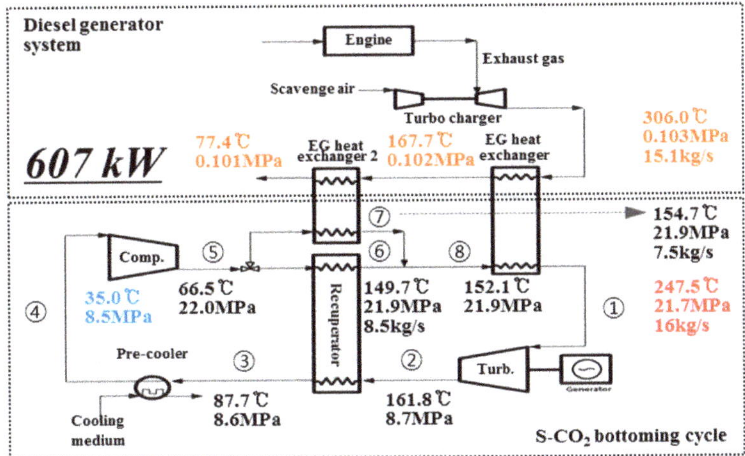

Figure 12. Configuration of partial heating cycle.

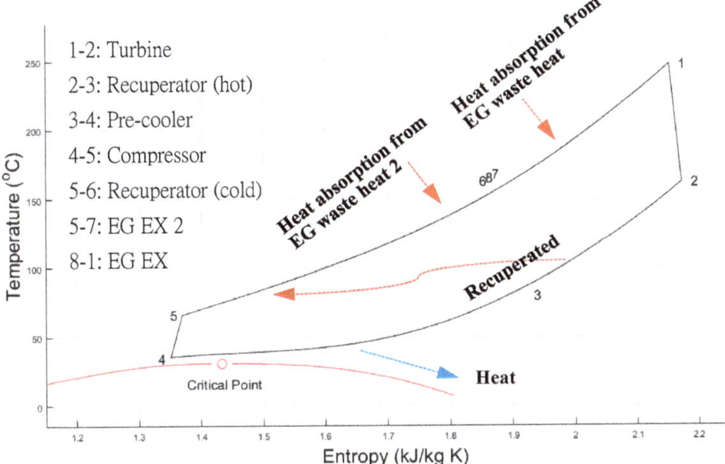

Figure 13. T-s diagram of the partial heating cycle.

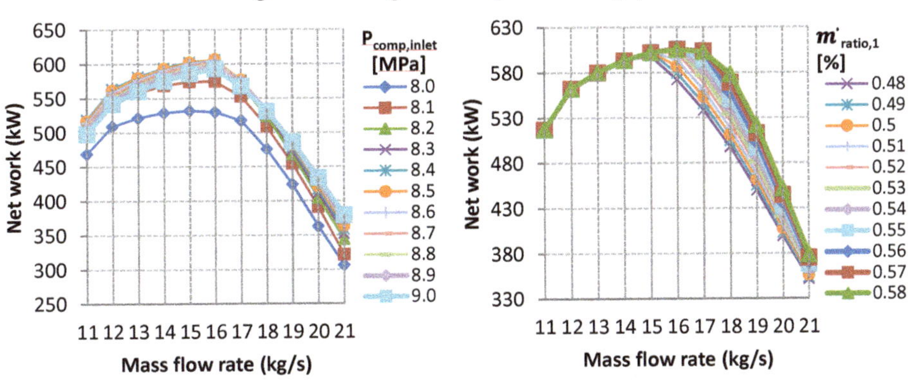

Figure 14. Net work dependence with mass flow rate, flow split, and turbine outlet pressure.

2.3.5. Partial Heating Cycle with SA Heat

The partial heating cycle with SA heat modified from partial heating cycle is analyzed to use the SA heat from the AAC DG turbocharger (see Figure 15) additionally. It has the same number of heat exchangers and turbomachinery with the partial heating cycle, but EG heat exchanger 2 was changed to SA heat exchanger for absorption of SA heat.

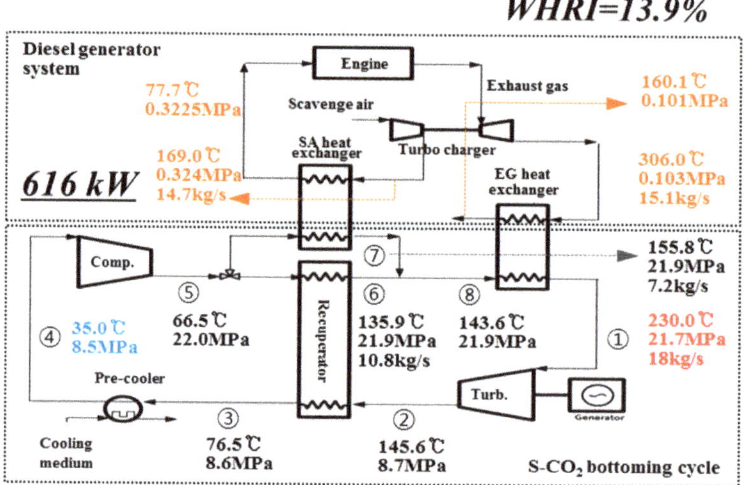

Figure 15. Configuration of partial heating cycle with SA heat.

The T-s diagram of the partial heating cycle with SA heat is shown in Figure 16. Figure 17 shows the net produced work variation with respect to the design variables. The partial heating cycle with SA heat has 13.9% WHRI and 616 kWe net produced work which is the 0.2% and 9 kWe (+1.5%) higher than the partial heating cycle. Unlike the simple recuperation cycle with SA heat, waste heat recovery is increased slightly from 3.63 MWth to 3.70 MWth (+1.9%) since the outlet temperature of EG heat exchanger is similar to the outlet temperature of scavenge air.

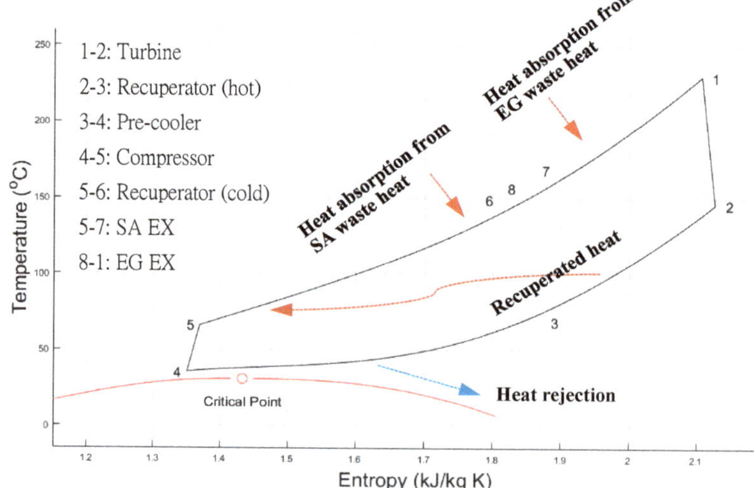

Figure 16. T-s diagram of the partial heating cycle with SA heat.

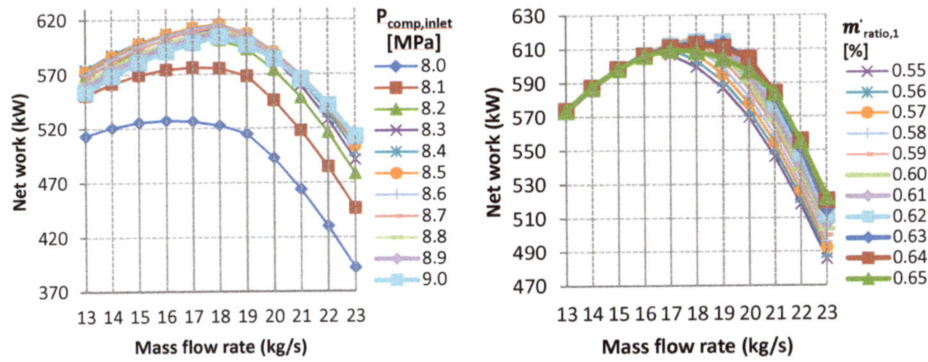

Figure 17. Net work dependence with mass flow rate, flow split, and turbine outlet pressure.

2.3.6. Modified Partial Heating Cycle with SA Heat

To increase the cycle performance, the partial heating cycle with SA heat was modified by changing the flow split merging point from cold side inlet to cold side outlet of EG heat exchanger (see Figure 18). It has the same number of heat exchangers and turbomachinery with the partial heating cycle with SA heat.

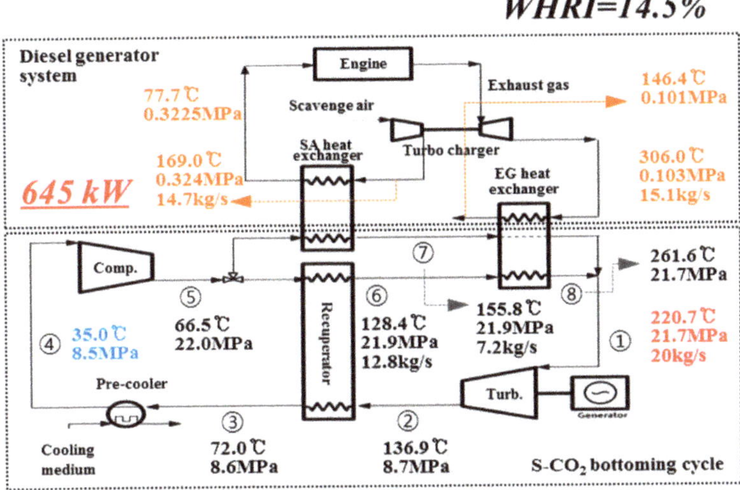

Figure 18. Configuration of modified partial heating cycle with SA heat.

The T-s diagram of the modified partial heating cycle with SA heat for the S-CO$_2$ bottoming cycle is shown in Figure 19. Figure 20 shows the net produced work dependence on design variables. The modified partial heating cycle with SA heat has 14.5% WHRI and 645 kWe net produced work which is the 0.8% and 38 kWe (+6.3%) higher than the partial heating cycle. By changing the flow split merging point from cold side inlet to cold side outlet of EG heat exchanger, waste heat recovery is increased from 3.63 MWth to 3.91 MWth (+7.7%) since the temperature of cold side inlet of EG heat exchanger is decreased.

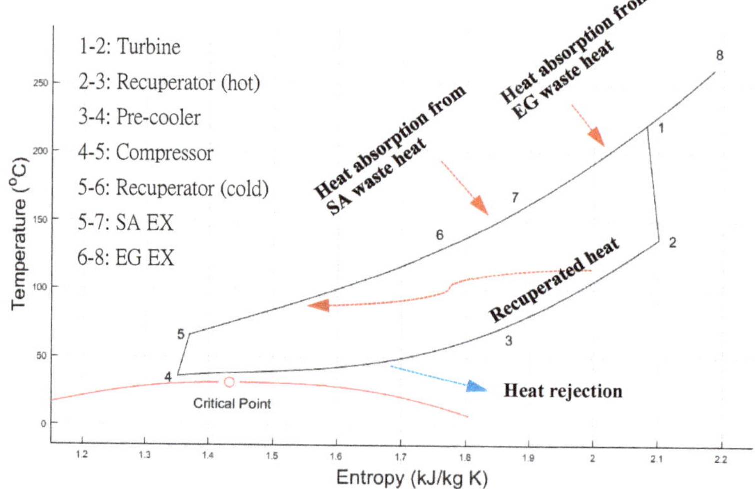

Figure 19. T-s diagram of the modified partial heating cycle with SA heat.

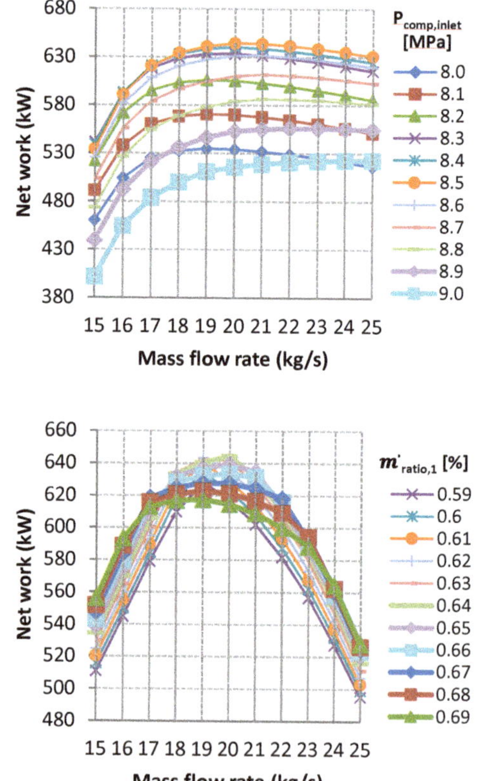

Figure 20. Net work dependence with mass flow rate, flow split, and turbine outlet pressure.

3. Discussion and Summary

Table 9 summarizes the main characteristics of various S-CO$_2$ cycles for a waste heat recovery of 7.2 MW AAC DG. It is clear from Table 9 that the recompression cycle is not the best cycle to recover waste heat of the DG due to the small temperature difference in the EG and SA heat exchangers. In this study, the net work of the modified partial heating cycle with SA heat achieved 645 kWe, which is the highest among the analyzed S-CO$_2$ cycles. Because emission factor considered for a diesel generator was 1.27 kg of CO$_2$/kWh [23], modified partial heating cycle with SA heat can reduce 819.2 kg of CO$_2$/h. Having lower values with optimal CO$_2$ mass flow rate, number of components, and total exchanged heat means that the selected cycle has more preferable characteristics compared to the others. Lower CO$_2$ mass flow rate has benefits in pipe cost and pressure drop. Also, lower number of components, and lower total exchanged heat have benefits in CAPEX and OPEX since the system volume is most critical design parameter [24]. Cycle net efficiency, net work and WHRI are the main concerns in system performance and the performance of the modified partial heating cycle with SA heat is to the best among the others. The optimal CO$_2$ mass flow rate of the modified partial heating cycle with SA heat is 20 kg/s which is larger than that of the exhaust gas.

Table 9. Main characteristics comparison with S-CO$_2$ cycles for a bottoming cycle.

Cycle Layout	Optimal CO$_2$ Mass Flow Rate (kg/s)	Number of HX/Turb./Comp. (-)	Total Exchanged Heat (MWth)	Waste Heat Recovery (MWth)	Cycle Net Efficiency (%)	Net Produced Work (kWe)	WHRI (%)
(1-1) Simple recuperated cycle	12	3/1/1	6.23	2.55	20.20	515	11.6
(1-2) Simple recuperation cycle with SA heat	16	4/1/1	7.58	3.21	17.67	568	12.8
(2-1) Recompression cycle with SA heat	18	5/1/2	8.29	3.45	17.65	613	13.8
(3-1) Partial heating cycle	16	4/1/1	8.17	3.63	16.73	607	13.7
(3-2) Partial heating cycle with SA heat	18	4/1/1	8.46	3.70	16.67	616	13.9
(3-3) Modified partial heating cycle with SA heat	20	4/1/1	8.99	3.91	16.47	**645**	**14.5**

4. Component Design

4.1. Turbomachinery Design

The turbomachinery was designed to arrange components optimally and to conduct the off-design analysis by predicting the performance for the modified partial heating cycle with SA heat. It is generally known that in case of small scale power conversion systems under a few tens of MW level, radial turbomachinery is suitable since the volumetric flow rate of working fluid is relatively low enough to use axial turbomachinery. Argonne National Laboratory (ANL) has suggested suitable design choices for designing S-CO$_2$ turbomachinery. The ANL also presented that small scale turbomachinery of S-CO$_2$ system is better to design with radial type [25]. Therefore, the radial type turbomachinery was selected for component design of the modified partial heating cycle with SA heat.

In this turbomachinery design study, the Balje's diagram known as the most useful turbomachinery preliminary design method was utilized for the design prediction [26]. The base parameters of turbomachinery design can be easily obtained with the specific speed (n$_s$) and diameter (d$_s$) since Balje's diagram suggests optimal values. Additionally, it represents the efficiency of turbomachinery depends on specific speed as shown in Figure 21. Table 10 shows the turbomachinery design results such as stages, impeller diameter and rotational speed for the designed operating condition by using the Balje's diagram with the non-dimensional numbers.

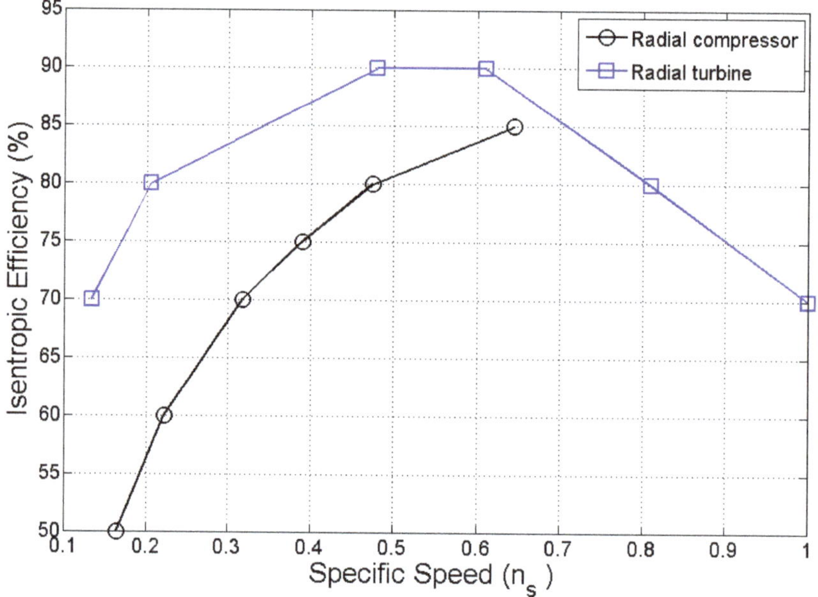

Figure 21. Efficiency variation with specific speed (recreation of figure in Ref. [26]).

Table 10. Turbomachinery design of the modified partial heating cycle with SA heat.

	Type: Radial						
	Mass flow rate (kg/s)	Stages	Diameter (m)	Specific Speed	Specific diameter	Efficiency (%)	RPM
Turbine	20	1	0.1	0.54	4.33	88	56,000
Compressor	20	1	0.07	0.63	4.62	75	56,000

The specific speed and diameter:

$$n_s = \frac{\omega \sqrt{V_1}}{\left(gH_{ad}\right)^{3/4}}, \quad d_s = \frac{D\left(gH_{ad}\right)^{1/4}}{\sqrt{V_1}} \tag{5}$$

The performance maps prediction was obtained with the in-house code KAIST-Turbomachinery Design (KAIST-TMD [27]. The turbomachinery design method of the KAIST-TMD code is a one dimensional mean-line analysis including a real gas approach for overcoming the limitation of conventional turbomachinery design tools near the S-CO$_2$ critical point. The reliability of the 1D turbomachinery code depends on choice of loss models. The loss model combination for radial compressor was validated with S-CO$_2$ compressor test data [28]. However, due to the lack of S-CO$_2$ turbine test data, the loss model combination for radial turbine was validated with test data of air radial turbine. Although S-CO$_2$ has a higher Reynolds number than air, it is acceptable to utilize air turbine data as an alternative, because the effect on turbine performance can be ignored over half a million in Reynolds number and S-CO$_2$ behaves like ideal gas in turbine operation range [29]. The performance map results such as pressure ratio and efficiency of the turbomachinery for the modified partial heating cycle with SA heat was calculated using the KAIST-TMD code and are shown in Figures 22–25.

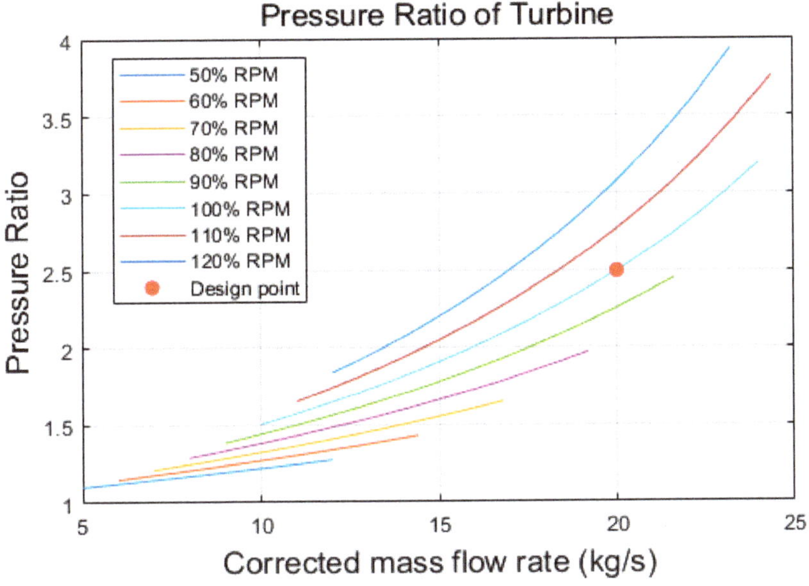

Figure 22. Turbine pressure ratio performance map of the modified partial heating cycle with SA heat.

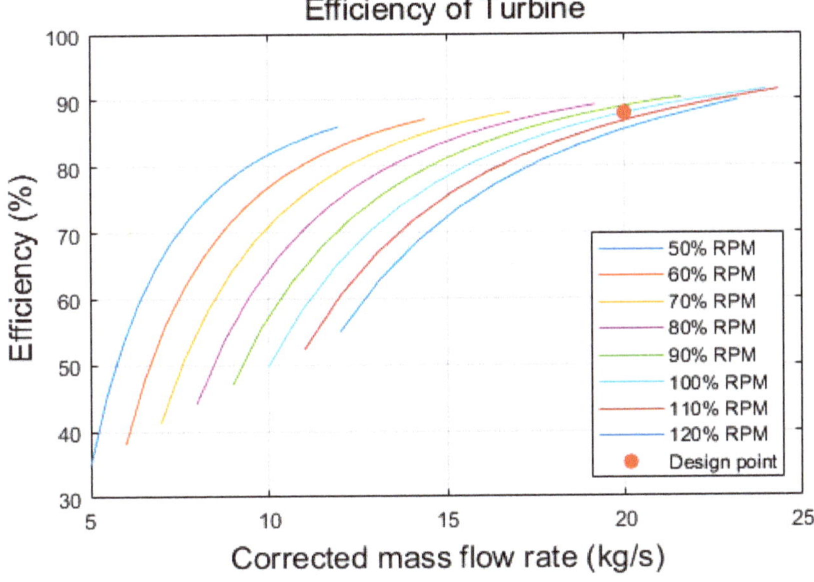

Figure 23. Turbine efficiency performance map of the modified partial heating cycle with SA heat.

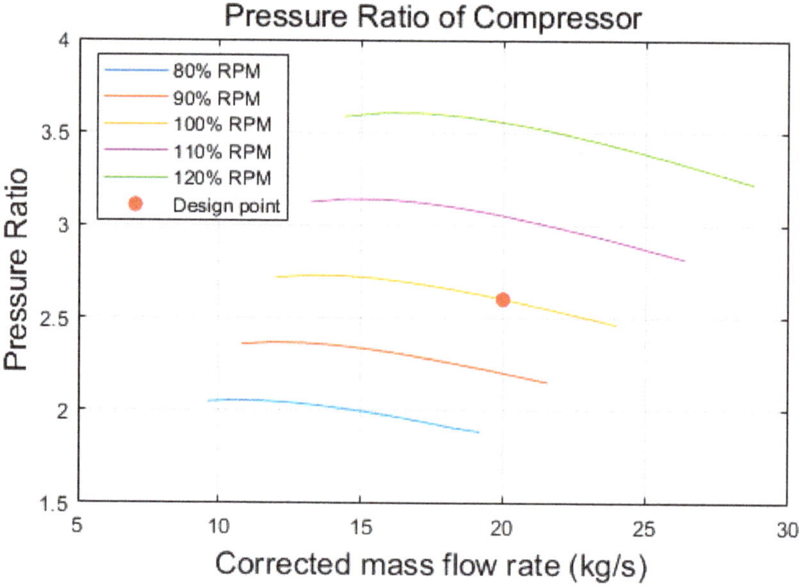

Figure 24. Compressor pressure ratio performance map of the modified partial heating cycle with SA heat.

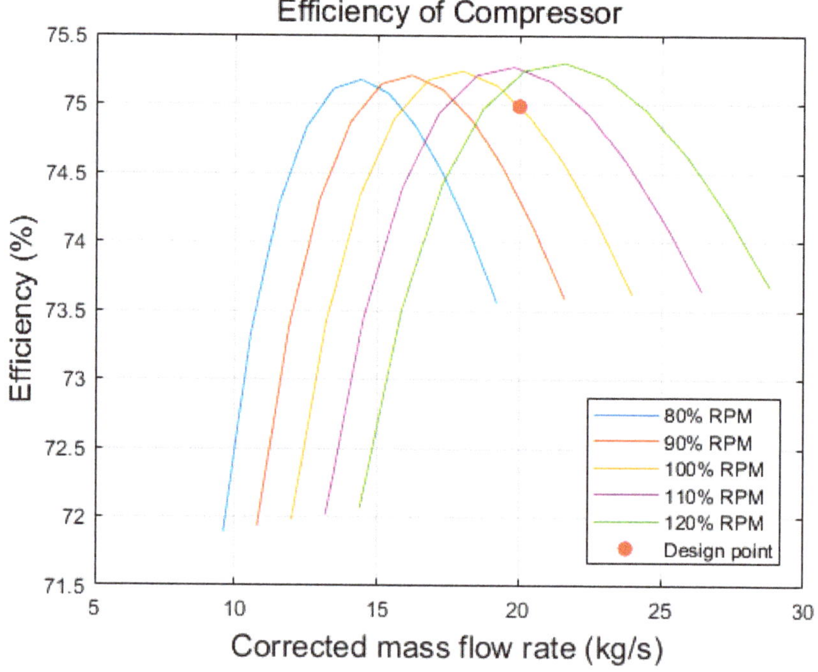

Figure 25. Compressor efficiency performance map of the modified partial heating cycle with SA heat.

4.2. Heat Exchanger Design

As previously mentioned in this paper, the main advantages of the S-CO$_2$ cycle are that it helps arrange the system in a compact way. It is well known that the overall footprint of the S-CO$_2$ system is directly related to the heat exchanger sizes among major components of the system because the turbomachinery size of the S-CO$_2$ system is significantly smaller than the heat exchanger size.

In this study, all heat exchangers of the modified partial heating cycle with SA heat are designed while assuming that the heat exchanger type is the Printed Circuit Heat Exchanger (PCHE). PCHE is selected to take full benefits such as high surface to volume density, wide operational range of pressure and temperature over 30 MPa and 800 °C, respectively [30].

In order to design heat exchangers of the cycle, an in-house PCHE design code, KAIST Heat eXchanger Design (KAIST-HXD) is used. Particularly the KAIST-HXD was developed to evaluate the geometry of the heat exchangers that meets requirements such as heat exchanger effectiveness and pressure drop values [31]. All the heat exchangers were designed with the configuration of a counter current flow and the optimal design conditions were also calculated to obtain the minimum volume while satisfying the target performance requirements. Table 11 summarizes the design result of the heat exchangers and Figure 26 shows the temperature profile in the recuperator.

Table 11. Heat exchanger design of the modified partial heating cycle with SA heat.

Parameter.	Pre-Cooler	Recuperator
Type	PCHE	PCHE
Shape	Zig-zag flow channel	Zig-zag flow channel
Hot channel fluid	CO$_2$	CO$_2$
Cold channel fluid	Water	CO$_2$
Hot channel D (semi-circular)	2.0 mm	2 mm
Cold channel D (semi-circular)	2.0 mm	1.8 mm
Hot channel No	8500	135,000
Cold channel No	8500	6750
Hot side Re # (Avg)	70,215	54,602
Cold side Re # (Avg)	5008	37,820
Length	0.78 m	0.96 m
Volume	0.088 m^3	0.12389 m^3
Pinch point	6.1 °C	4.97 °C

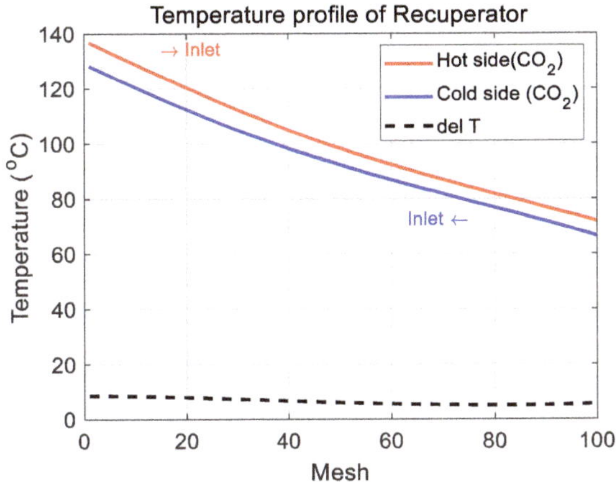

Figure 26. Temperature profile of the designed recuperator.

5. Off-Design Analysis

Quasi-Steady State Analysis

To establish the control logic for the part-load operation, an in-house quasi-static cycle analysis code, KAIST-QCD (Quasi-static Cycle Design) was developed by a KAIST research team [32]. In the system performance analysis, various factors influence the system behavior. Quasi-steady state is defined as a pseudo steady state of the system and the analysis method is applied when the off-design performance of the system is important but the transient performance is less of an interest. The quasi-steady state analysis provides quick and simple solutions that describe each operating condition in the system. Compared to the transient analysis, the quasi-steady state analysis requires less calculation time and provides useful information for designing control schemes for the system.

In this paper, the component design parameters of compressors, turbines and, heat exchangers described previously are used for the quasi-steady state analysis. During the quasi-steady state analysis, compressor inlet condition is first defined. Compressor performance is affected by two factors: the compressor inlet condition and the rotating speed. Based on turbine and compressor off-design performance maps, mass flow in the system is calculated. After that, heat loads between hot and cold side of heat exchangers are calculated. The algorithm of quasi-steady state analysis is shown in Figure 27. In this scenario, it is assumed that only mass inventory control was applied, and RPM was kept at design value for grid stability.

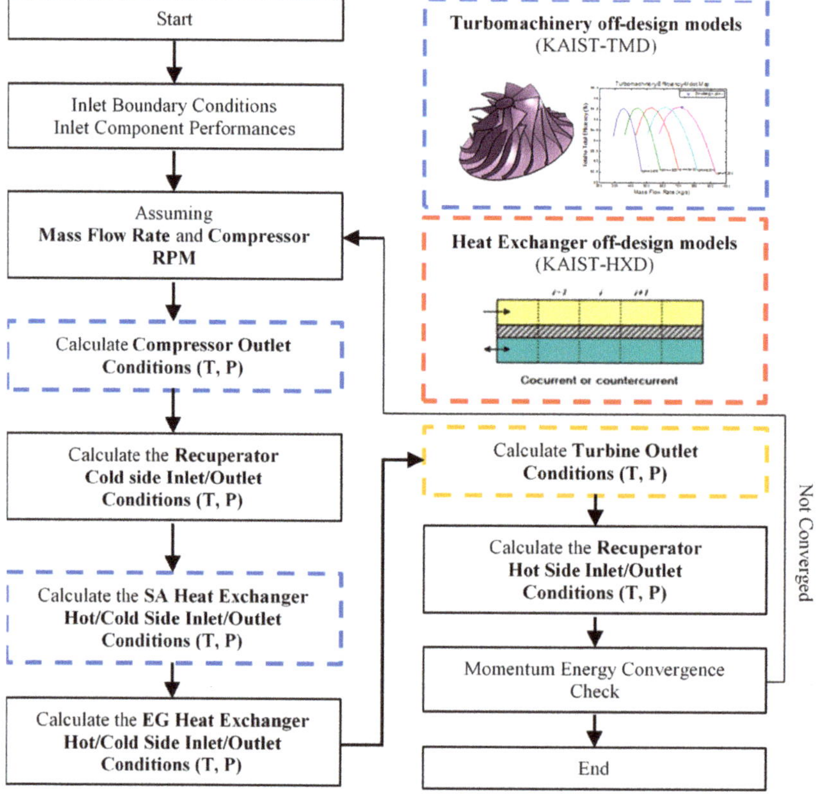

Figure 27. S-CO$_2$ cycle quasi-steady state analysis code algorithm.

Off-design performance of a waste heat recovery system at 63, 94, and 100% thermal load, respectively, is shown in Figures 28–30. As shown in figures, as the load increases, the optimum points are formed at larger system flow rates and larger flow split ratio to the SA heat exchanger ($\dot{m}_{\mathrm{ratio}}$). This means that the lower the thermal load, the higher the performance can be achieved when controlled to reduce the flow to the SA heat exchanger. These results show an output difference of as much as 100 kW or more between the optimal and non-optimal points. Therefore, it is necessary to further study the strategy of controlling the driving point to be followed by producing the optimal driving point according to each off-design operation mode in advance. The performance change of off-design optimum point according to thermal load is shown in Figure 31. A loss of 6.74 kW net-work is expected for 1%p thermal load decrease.

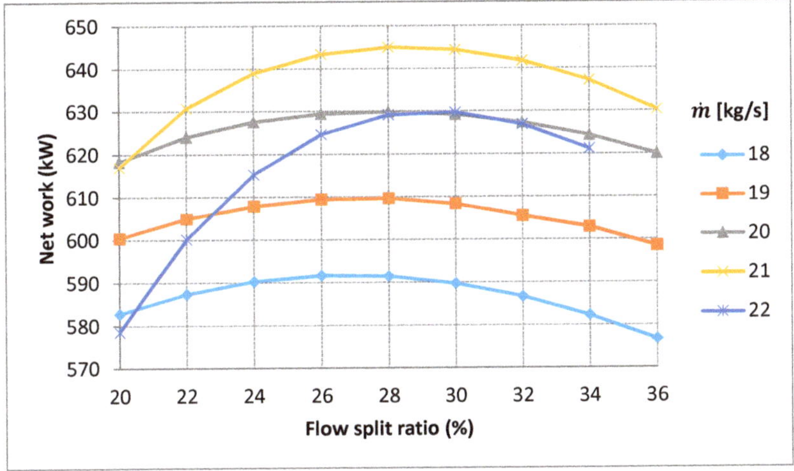

Figure 28. Quasi-steady state analysis result (100% load).

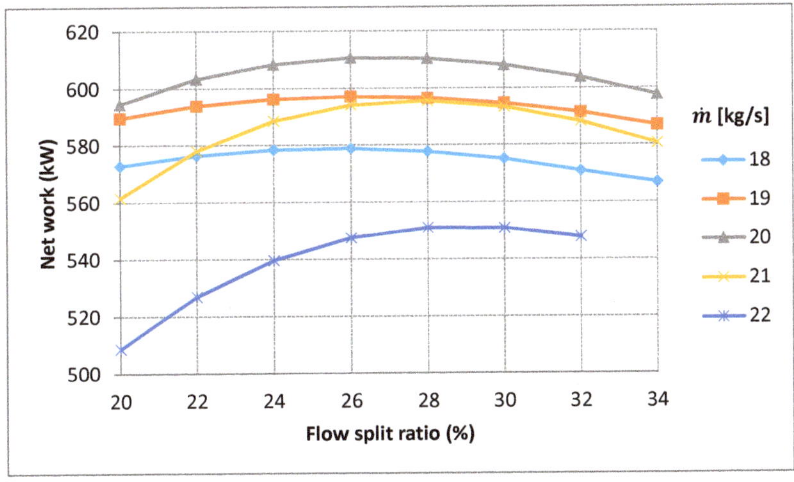

Figure 29. Quasi-steady state analysis result (94% load).

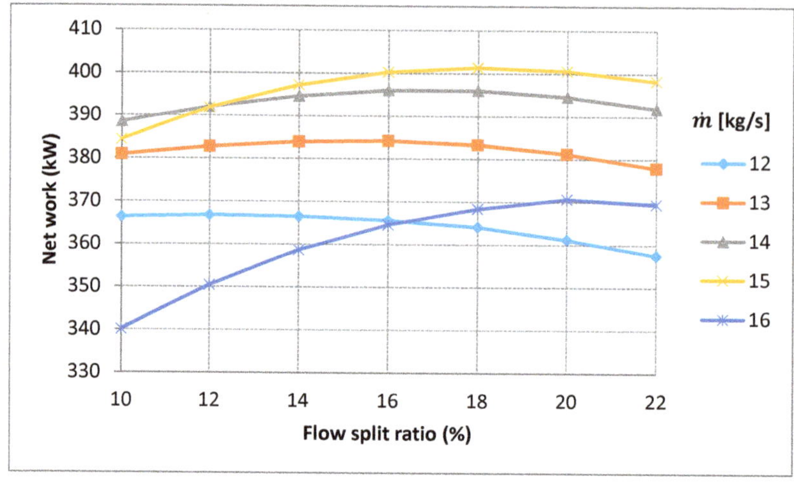

Figure 30. Quasi-steady state analysis result (63% load).

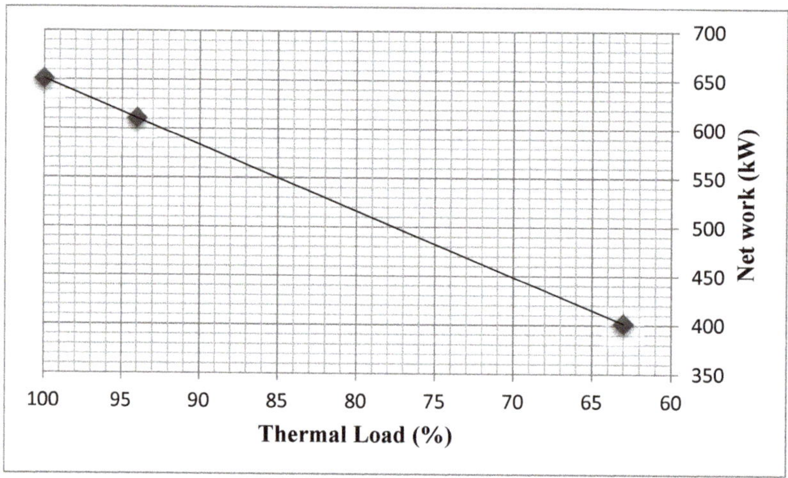

Figure 31. S-CO$_2$ cycle performance in part load conditions.

6. Conclusions

In this paper, results of the thermodynamic cycle analyses and sensitivity analyses of various S-CO$_2$ cycles for the waste heat recovery of a 7.2 MW diesel engine used in a nuclear power plant are presented. The diesel engine can provide two heat sources for the waste heat recovery system: one from exhaust gas and the other from scavenge air. This is unique compared to other heat sources such as fossil fuel power plants, concentrated solar power systems, nuclear power plants, etc. Thus, it was necessary to review the S-CO$_2$ cycle layouts for dual heat sources application.

As a result of cycle comparisons, it was found that the partial heating cycle produces the highest net work while having a simple layout and small number of major components. Particularly, the partial heating cycle has additional advantage in respect of easier cycle operation scheme since it requires single turbine and compressor only.

Furthermore, the partial heating cycle was modified by changing the flow split merging point from cold side inlet to cold side outlet of exhaust gas heat exchanger, to additionally use the scavenge air heat from diesel engine turbocharger. The waste heat recovery is sharply increased by applying dual heat sources from exhaust gas and scavenge air. This study identified that waste heat recovery can be increased more by changing the flow split merging point of a partial heating cycle for the diesel engine application.

The main components, including turbomachinery and heat exchangers, are designed with in-house codes which have been validated with experiment data. Based on the designed cycle and components, the S-CO_2 cycle performance under partial load conditions is analyzed with an in-house quasi-steady state cycle analysis code to find the optimal operating condition and to establish the control logic for the part-load operation. In this paper, part-load operation and optimal operating conditions were calculated varying from 63–100%. For the off-design analysis it is performed while assuming only mass inventory control is applied. RPM was kept at design value for the grid stability. Part load analysis shows an output difference of as much as 100 kW or more between the optimal and non-optimal points. Therefore, it is necessary to further study the strategy of controlling the operation point for each off-design operation mode. Moreover, the performance change of off-design optimum point for varying thermal load is important for economic analysis of the system.

The additional power from the AAC DG waste heat recovery system can operate a nuclear safety system without revising the existing nuclear power plant's layout too much. By producing additional 645 kWe from the S-CO_2 bottoming cycle from 3 MWth waste heat, nuclear safety can be improved by potentially extending AAC DG operation time or adopting more safety components to improve redundancy or diversity. Furthermore, more instrument and control systems can be utilized during accident from the additional power of the waste heat recovery system.

Author Contributions: Conceptualization, J.K.H.; Methodology, J.K.H. and M.S.K.; Software, J.K.H., M.S.K. and S.S.; Validation, J.K.H., M.S.K. and J.L.; Formal Analysis, J.K.H. and M.S.K.; Investigation, B.S.O.; Resources, J.L.; Data Curation, S.S.; Writing-Original Draft Preparation, J.K.H.; Writing-Review & Editing, J.I.L.; Visualization, M.S.K.; Supervision, J.I.L.; Project Administration, J.I.L.; Funding Acquisition, J.I.L.

Funding: This work was supported by the National Research Foundation of Korea (NRF) grant funded by the Korea government (MSIP) (2017M2B2B1071971).

Acknowledgments: This work was supported by the National Research Foundation of Korea (NRF) grant funded by the Korea government (MSIP) (2017M2B2B1071971).

Conflicts of Interest: The authors declare no conflict of interest.

Nomenclature

η:	efficiency
$\dot{m}$:	mass flow rate
h:	enthalpy
T:	temperature
P:	pressure
s:	entropy
W:	work
$P_{comp,\ inlet}$:	compressor inlet pressure
$\dot{m}_{ratio}$:	flow split ratio
c_p:	specific heat capacity
ρ:	fluid density
ω:	the angular frequency of the rotor
V:	velocity
g:	gravity force acceleration
H:	head
D:	diameter
n_s:	specific speed
d_s:	specific diameter

Acronyms

S-CO$_2$:	supercritical carbon dioxide
MW:	mega watt
AAC DG:	alternate alternating current diesel generator
SBO:	station black out
EDG:	emergency diesel generator
NPPs:	nuclear power plants
APR1400:	advanced power reactor with an electrical power output of 1400 MW
WHRI:	waste heat recovery index
ORC:	organic Rankine cycle
KAIST-CCD:	KAIST-closed cycle design
REFPROP:	reference fluid thermodynamic and transport properties database
MT:	main turbine
MC:	main compressor
HT:	high temperature
LT:	low temperature
SA:	scavenge air
C:	compressor
PC:	precooler
HX:	heat exchanger
TB:	turbine
CP:	compressor
ANL:	argonne national laboratory
KAIST-TMD:	KAIST-turbomachinery design
1D:	one-dimension
KAIST-HXD:	KAIST-heat exchanger design

References

1. Kim, B.K.; Jeong, B.S.; Jo, M.K. *Post-Fukushima Action Plan in Korea*; Oxford Christ Church: Oxford, UK, 2011.
2. Emirates Nuclear Energy Corporation. *BRAKA Nuclear Power Plant Units 1&2 Preliminary Safety Analysis Report*; Emirates Nuclear Energy Corporation: Abu Dhabi, UAE, 2010.
3. Lee, S.K. Development Status of Emergency Diesel Generator Status Diagnosis in Nuclear Power Plants. *J. Electr. World* **2014**, *Special And+*, 82–88.
4. Dostal, V.; Hejzlar, P.; Driscoll, M.J. The supercritical carbon dioxide power cycle: Comparison to other advanced power cycles. *Nucl. Technol.* **2006**, *154*, 283–301. [CrossRef]
5. Chen, Y.; Lundqvist, P.; Johansson, A.; Platell, P. A comparative study of the carbon dioxide transcritical power cycle compared with an organic rankine cycle with R123 as working fluid in waste heat recovery. *Appl. Therm. Eng.* **2006**, *26*, 2142–2147. [CrossRef]
6. Yoon, H.J.; Ahn, Y.; Lee, J.I.; Addad, Y. Potential advantages of coupling supercritical CO$_2$ Brayton cycle to water cooled small and medium size reactor. *Nucl. Eng. Des.* **2012**, *245*, 223–232. [CrossRef]
7. Feher, E.G. Supercritical thermodynamic power cycle, presented to the IECEC. *Douglas Paper*, 2 August 1967; p. 4348.
8. Bae, S.J.; Lee, J.; Ahn, Y.; Lee, J.I. Preliminary studies of compact Brayton cycle performance for small modular high temperature gas-cooled reactor system. *Ann. Nucl. Energy* **2015**, *75*, 11–19. [CrossRef]
9. Martelli, E.; Nord, L.; Bolland, O. Design criteria and optimization of heat recovery steam cycles for integrated reforming combined cycles with CO$_2$ capture. *Appl. Energy* **2012**, *92*, 255–268. [CrossRef]
10. Heo, J.Y.; Kim, M.S.; Baik, S.; Bae, S.J.; Lee, J.I. Thermodynamic study of supercritical CO$_2$ Brayton cycle using an isothermal compressor. *Appl. Energy* **2017**, *206*, 1118–1130. [CrossRef]
11. Ahn, Y.; Lee, J.; Kim, S.G.; Lee, J.I.; Cha, J.E. The design study of supercriticalcarbon dioxide integral experiment loop. In Proceedings of the ASME Turbo Expo 2013: Turbine Technical Conference and Exposition, San Antonio, TX, USA, 3–7 June 2013.

12. Cho, S.K.; Kim, M.; Baik, S.; Ahn, Y.; Lee, J.I. Investigation of the bottoming cycle for high efficiency combined cycle gas turbine system with supercritical carbon dioxide power cycle. In Proceedings of the ASME Turbo Expo 2015: Turbine Technical Conference and Exposition, Montreal, QC, Canada, 15–19 June 2015.

13. Kim, M.S.; Ahn, Y.; Kim, B.; Lee, J.I. Study on the supercritical CO_2 power cycles for landfill gas firing gas turbine bottoming cycle. *Energy* **2016**, *111*, 893–909. [CrossRef]

14. Lemmon, E.W.; Huber, M.L.; McLinden, M.O. *NIST Standard Reference Database 23: NIST Reference fluid Thermodynamic and Transport Properties-REFPROP, Version 9.0*; National Institute of Standards and Technology: Gaithersburg, DC, USA, 2010.

15. Pavri, R.; Moore, G.D. *Gas Turbine Emissions and Control*; GE Power Systems: Atlanta, GA, USA, 2001.

16. Cziesla, F.; Bewerunge, J.; Senzel, A. Lünen e state-of-the art ultra supercritical steam power plant under construction. In Proceedings of the POWER-GEN Europe 2009, Cologen, Germany, 26–29 May 2009.

17. Kimzey, G. Development of a brayton bottoming cycle using supercritical carbon dioxide as the working fluid, a report submitted in partial fulfillment of the requirements for gas turbine industrial fellowship. In *University Turbine Systems Research Program*; Electric Power Research Institute: Palo Alto, CA, USA, 2012.

18. Musgrove, G.O.; Pittaway, C.; Vollnogle, E.; Chordia, L. Tutorial: Heat exchangers for supercritical CO_2 power cycle applications. In Proceedings of the 5th International Symposium-Supercritical CO_2 Power Cycles, San Antonio, TX, USA, 28–31 March 2016.

19. Bae, S.J.; Lee, J.I.; Ahn, Y.H.; Lee, J.K. Various supercritical carbon dioxide cycle layouts study for molten carbonate fuel cell application. *J. Power Sources* **2014**, *270*, 608–618. [CrossRef]

20. Ahn, Y.H.; Lee, J.I. Study of various brayton cycle designs for small modular sodium-cooled fast reactor. *Nucl. Eng. Des.* **2014**, *276*, 128–141. [CrossRef]

21. Kulhanek, M.; Dostal, V. Thermodynamic analysis and comparison of supercritical carbon dioxide cycles. In Proceedings of the Supercritical CO_2 Power Cycle Symposium, Boulder, CO, USA, 24–25 May 2011.

22. Campanari, S.; Macchi, E. Thermodynamic analysis of advanced power cycles based upon solid oxide fuel cells, gas turbines and rankine bottoming cycles, 98-GT-585. In Proceedings of the ASME International Gas Turbine and Aero engine Congress and Exhibition, Stockholm, Sweden, 2–5 June 1998.

23. Jakhrani, A.Q.; Othman, A.; Rigit, A.R.H.; Samo, S.R. Estimation of carbon footprints from diesel generator emissions. In Proceedings of the International Conference Green Ubiquitous Technology, Jakarta, Indonesia, 7–8 July 2012.

24. Hewitt, C.H.; Pugh, S.J. Approximate design and costing methods for heat exchangers. *Heat Transf. Eng.* **2007**, *28*, 76–86. [CrossRef]

25. Sienicki, J.J.; Moisseytsev, A.; Fuller, R.; Wright, S.; Pickard, P. Scale Dependencies of Supercritical Carbon Dioxide Brayton Cycle Technologies and Optimal Size for a Next-Step Supercritical CO_2 Cycle Demonstration. In Proceedings of the Supercritical CO_2 Power Cycle Symposium, Boulder, CO, USA, 24–25 May 2011.

26. Balje, O.E. *Turbomachines: A Guide to Design, Selection, and Theory*; A Wiley-Interscience Publication: New York, NY, USA, 1980.

27. Lee, J.; Lee, J.I.; Yoon, H.J.; Jae, E.C. Supercritical Carbon Dioxide turbomachinery design for water-cooled Small Modular Reactor application. *Nucl. Eng. Des.* **2017**, *270*, 76–89. [CrossRef]

28. Lee, J.; Cho, S.; Lee, J.I. The Effect of Real Gas Approximations on S-CO_2 Compressor Design. *J. Turbomach.* **2018**, *140*. [CrossRef]

29. Cho, S.K.; Lee, J.; Lee, J.I.; Cha, J.E. S-CO_2 turbine design for decay heat removal system of sodium cooled fast reactor. In Proceedings of the ASME Turbo Expo 2016: Turbomachinery Technical Conference and Exposition, Seoul, Korea, 13–17 June 2016.

30. Shah, R.K.; Sekulic, D.P. *Fundamentals of Heat Exchanger*; John Wiley & Sons, Inc.: Hoboken, NY, USA, 2002.

31. Baik, S.; Kim, S.G.; Lee, J.; Lee, J.I. Study on CO_2-water printed circuit heat exchanger performance operating under various CO_2 phases for S-CO_2 power cycle application. *Appl. Therm. Eng.* **2017**, *113*, 1536–1546. [CrossRef]

32. Ahn, Y.H.; Kim, M.S.; Lee, J.I. S-CO_2 cycle design and control strategy for the SFR application. In Proceedings of the 5th International Symposium–Supercritical CO_2 Power Cycles, San Antonio, TX, USA, 28–31 March 2016.

Article

Potential of Supercritical Carbon Dioxide Power Cycles to Reduce the Levelised Cost of Electricity of Contemporary Concentrated Solar Power Plants

Francesco Crespi, David Sánchez *, Gonzalo S. Martínez and Tomás Sánchez-Lencero and Francisco Jiménez-Espadafor

Department of Energy Engineering, University of Seville, Camino de los Descubrimientos s/n, 41092 Seville, Spain; crespi@us.es (F.C.); gsm@us.es (G.S.M.); tmsl@us.es (T.S.-L.); fcojjea@us.es (F.J-E.)
* Correspondence: ds@us.es; Tel.: +34-954-486-488

Received: 31 May 2020; Accepted: 15 July 2020; Published: 22 July 2020

Featured Application: Assessment of the potential of supercritical Carbon Dioxide power cycles to reduce the Levelised Cost of Electricity of contemporary Concentrated Solar Power plants, with the aim to benchmark the cost of electricity of the current and next generation Concentrated Solar Power (CSP) technology.

Abstract: This paper provides an assessment of the expected Levelised Cost of Electricity enabled by Concentrated Solar Power plants based on Supercritical Carbon Dioxide (sCO$_2$) technology. A global approach is presented, relying on previous results by the authors in order to ascertain whether these innovative power cycles have the potential to achieve the very low costs of electricity reported in the literature. From a previous thermodynamic analysis of sCO$_2$ cycles, three layouts are shortlisted and their installation costs are compared prior to assessing the corresponding cost of electricity. Amongst them, the *Transcritical* layout is then discarded due to the virtually impossible implementation in locations with high ambient temperature. The remaining layouts, Allam and Partial Cooling are then modelled and their Levelised Cost of Electricity is calculated for a number of cases and two different locations in North America. Each case is characterised by a different dispatch control scheme and set of financial assumptions. A Concentrated Solar Power plant based on steam turbine technology is also added to the assessment for the sake of comparison. The analysis yields electricity costs varying in the range from 8 to over 11 ¢/kWh, which is near but definitely not below the 6 ¢/kWh target set forth by different administrations. Nevertheless, in spite of the results, a review of the conservative assumptions adopted in the analysis suggests that attaining costs substantially lower than this is very likely. In other words, the results presented in this paper can be taken as an upper limit of the economic performance attainable by Supercritical Carbon Dioxide in Concentrated Solar Power applications.

Keywords: LCoE; CSP; supercritical CO$_2$

---

1. Introduction

1.1. Current Status of Concentrated Solar Power Technology. Expectations Raised by Supercritical Carbon Dioxide Power Cycles

Supercritical Carbon Dioxide power cycles are currently seen as the technology of choice for next generation Concentrated Solar Power plants to produce Solar Thermal Electricity (STE). They promise higher efficiencies than state-of-the-art Concentrated Solar Power (CSP)-STE facilities based on steam technology, and they are reportedly much more compact thanks to the higher density of the working fluid and the simpler layout of the working cycle. Nevertheless, this latter feature, compactness of the power block, does not constitute an essential difference with respect to conventional steam technology, given that the power block contributes a fairly small fraction to the land area occupied by a Concentrated Solar Power plant. On the contrary, the higher efficiency of sCO_2 systems (as compared to contemporary technology) reduces the heat input needed for a given power output and thermal energy storage capacity, which translates into a smaller solar field and TES system. This turns out to be crucial for reducing both the footprint and cost of this new generation of CSP-STE systems based on sCO_2.

In 2011, the SunShot programme, flagship R&D instrument of the United States Department of Energy to foster the development of solar power generation, set the objective to reduce the cost of solar electricity to 6 ¢/kWh in 2020. This ambitious objective was already achieved in 2017 by photovoltaic technology in large-scale facilities, and it has been superseded by a much more ambitious goal of 3 ¢/kWh in 2030. Unfortunately, whilst PV is now cost-effective and competitive against other renewable and non-renewable technologies, even in smaller domestic applications, it looks like CSP-STE is experiencing more difficulties to meet these objectives with the technology currently available. This is where the cumulative cost reduction potential brought about by sCO_2 power cycles in combination with more mature solar field and thermal energy storage technologies, along with the exploitation of economies of scale of the CSP industry, comes in to pave the way for a drastic cost reduction of Solar Thermal Electricity. Figure 1, adapted from the information made public by Solar Energy Technologies Office of the Department of Energy [1], provides this roadmap for CSP-STE technologies.

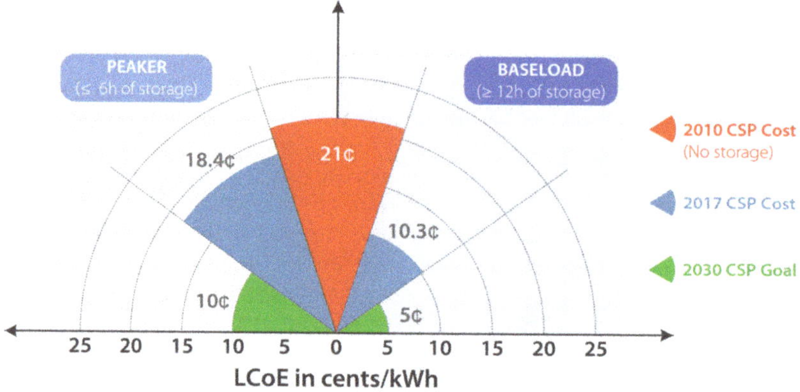

Figure 1. Progress and goals for Solar Thermal Electricity set by the SunShot programme. Data obtained from [1].

1.2. Technical Hurdles Hindering the Development of Supercritical Carbon Dioxide Power Cycles for CSP Applications

Unfortunately, even if the advent of supercritical Carbon Dioxide technologies in the beginning of the twenty-first century was seen as the perfect opportunity for a drastic cost reduction of CSP-STE, this perception has not materialised. The large research efforts needed to develop the technology, hence reducing the associated capital cost, and the much slower pace at which CSP is being deployed to the market, are hindering the construction of a first commercial (or even pre-commercial) plant demonstrating the concept. Moreover, in recent years, much more attention has been put on a seemingly unsolvable problem in the low temperature section of the sCO_2 cycle, which is the thermodynamic cornerstone of the reportedly better performance of this cycle with respect to conventional steam cycles.

The realisation of efficient supercritical CO_2 power cycles requires that the inlet temperature to the compressor be close to, or even lower than, the critical temperature of this fluid ($\approx$31 °C). When this is possible, and if the peak temperature of the cycle is higher than 600–650 °C, then the sCO_2 cycle outperforms any Rankine cycle running on water/steam with the same boundary conditions. Unfortunately, CSP plants are typically located in arid sites with ambient temperatures well above 35 °C, which makes it impossible to cool the cycle down to the temperatures needed to compress the fluid in the vicinity of the critical point (there where density is very high) with low power requirements. Accordingly, the rapid transition to an almost ideal behaviour of Carbon Dioxide when temperature increases to 40 °C or above increases compression work and reduces the thermal efficiency of the power block, which can only be increased again through a large increase of turbine inlet temperature. Of course, this poses new challenges in the area of receiver technology, heat transfer fluids and materials, and also brings about higher costs.

In order to overcome these thermodynamic problems, the SCARABEUS formulates a new conceptual approach to sCO_2 cycles whereby the composition of the working fluid is tailored to the high ambient temperatures typically found in CSP sites. This is enabled by the addition of certain dopants with higher critical temperature than Carbon Dioxide to the raw CO_2 used in standard sCO_2 cycles. The addition of these dopants increases the critical temperature of the mixture and enables liquid-like compression (in supercritical conditions) even at ambient temperatures as high as 40–45 °C or even 70 °C. The concept has already been formulated by partners of the consortium in [2,3], showing promising results, but it is too early to assess the impact of this new technology on the economic performance of CSP-STE. New equipment (turbomachinery and heat exchangers) is needed and the properties and thermal stability of the dopants considered must be confirmed in long-term operation at high temperature. This falls within the scope of the SCARABEUS project and more information will be presented in the near future.

1.3. Objectives and Novelty. Benchmarking the First Generation of CSP-sCO₂ Power Plants

Akin to what was done several years ago to benchmark the then innovative supercritical Carbon Dioxide cycles in Concentrated Solar Power applications, a new reference is now needed to understand whether or not the SCARABEUS technology will eventually yield lower Levelised Cost of Electricity than standard sCO_2 power cycles. This encapsulates the twofold objective of the present work, in brief: (i) to produce accurate estimates of the Levelised Cost of Electricity that is attainable for Concentrated Solar Power plants using supercritical CO_2 cycles; and (ii) to provide a benchmark for the innovative SCARABEUS technology currently under development.

In this regard, the authors of this paper have carried out an ambitious research to assess the true potential of sCO_2 power cycles, with the aim to provide a structured pathway for the thermo-economic feasibility analysis of this technology when applied to CSP power plants. The results of this research have been published regularly in the last four years, and the present paper represents the last, wrap-up piece of the process. The research starts with a very thorough review of the technical (scientific and industrial) works dealing with sCO_2 technologies [4]. Since the early works by Sulzer, Angelino and Feher [5–7], this work provides a comprehensive classification of virtually all the cycle layouts found in the literature with the aim to facilitate the comparison between different layouts. Amongst all these cycles, the twelve candidates fitting best into CSP applications are shortlisted for a rigorous and systematic thermodynamic analysis in [8], with the goal to assess their true potential, free from the inherent technical constraints brought about by contemporary technology-related limitations. The thermodynamic assessment presented in [8] is complemented by two additional papers by the authors, [9,10], aimed at assessing the expected component cost of each major equipment in the plant in order to estimate the installation costs of CSP plants using sCO_2 power cycles. Out of this analysis, two of the twelve configurations are found to provide the best thermal and economic performance: Allam and Partial Cooling. For these two layouts, dedicated models of performance to assess the corresponding off-design performance are developed with the aim to calculate the annual production of electricity (annual yield) for a particular location and set of boundary conditions [11].

With all the information described above, this paper constitutes the last step in this research, leading to the original objective of estimating the Levelised Cost of Electricity f CSP-sCO$_2$ power plants. As said before, this is the Key Performance Indicator needed to assess whether or not standard sCO$_2$ cycles or the new SCARABEUS concept will enable meeting the ambitious LCoE objective set forth by the SunShot programme. The value of the present paper lies on two main features. First, the wide portfolio of cycle layouts as produced by the review paper written by the authors in 2017 based on 160 references [4], which ensures that all the candidate cycles are included in the comparison. Second, the fact that the series of papers listed in the previous paragraphs are all consistent in the methodologies used and assumptions made. Indeed, each piece of research is built upon a previous one by the authors and, therefore, the thermodynamic references and environmental conditions are all consistent, the time value of money is taken into account to normalise currency, the on and off-design models of performance used to calculate the yield of each cycle are the same, and there is no bias in the comparative analysis of results. This applies to the entire process, from literature review to LCoE calculation.

2. Techno-Economic Assessment of CSP Based on sCO$_2$

2.1. Operating Conditions

The initial works on supercritical power cycles acknowledged the superior thermal performance of this technology, with respect to standard steam cycles, when turbine inlet temperatures are higher than 600/650 °C [6]. Below this value, the little gain (if any) brought about by sCO$_2$ cycles does not pay off the lower reliability and higher cost of a still less mature technology. Unfortunately, even at the lower end of the temperature range where sCO$_2$ is substantially better than steam, the operating temperatures that are needed are far from the values currently achieved by commercial solar technologies, as recently confirmed by a report issued by the SCARABEUS consortium [12]. Two major hurdles are identified. Receiver technologies able to achieve temperatures in excess of 700 °C are available but they have not reached the commercial stage yet [13]. This also applies to heat transfer fluids able to harvest this high-temperature thermal energy available in the receiver, in order to feed both the power block and Thermal Energy Storage system, which are currently available but not widely commercialised [14]. This availability of high temperature heat transfer fluids is assessed in Table 1 where the characteristics of different intermediate to high-temperature molten salts are listed. Amongst them, this work makes use of FLiNaK for the sake of continuity with past works by the authors. Thermodynamic-wise, this is a very interesting option even though the very high cost of this salt compromises the economic feasibility of its practical implementation. Still, it is assumed that adopting this fluid yields a safe (conservative) estimate of the cost of electricity associated to this technology.

Table 1. Comparison between different molten salts for intermediate-high temperature Concentrated Solar Power (CSP) plants.

Salt	Composition [%]	Freezing Point [°C]	Boiling Point [°C]	Price [\$/kg]	Price [\$/kWh$_t$] *
NaNO$_3$-KNO$_3$ (Solar Salt) [15]	60–40	220	600	0.8	10
LiF-NaF-KF (FLiNaK) [14]	46.5–11.5–42	454	1570	8.6 [16]	54.8
Li$_2$CO$_3$-Na$_2$CO$_3$-K$_2$CO$_3$ [17]	32–33–35	397	662	2.5 [15]	26.1
Na$_2$CO$_3$-NaOH [17]	19–81	284	714	-	2.3
MgCl$_2$-KCl [15]	37.5–62.5	426	1412	0.35	5
ZnCl-NaCl-KCl [15]	69–7–24	204	732	0.8	18

* Values taken from [15] are obtained for ΔT_{salt} = 200 °C. For the FLiNaK, authors calculated this price for ΔT_{salt} = 290 °C. Values taken from [17] are estimated with market price of raw components in Q1 2015, for a maximum cycle temperature of 650 °C.

The thermodynamic potential of sCO$_2$ power cycles for different boundary conditions and applications was assessed by the authors in a previous work [8], based on an earlier, thorough literature review [4]. In a second step, turbine inlet temperature was set to 750 °C, based on several works in literature [18,19] and on private communications with Abengoa (a SCARABEUS partner), and the assessment of thermal performance was complemented by an economic analysis with the aim to calculate the capital cost of a representative CSP plant using each cycle layout [10]. The main specifications of this reference plant are presented in Table 2.

Table 2. Specifications of the reference sCO$_2$ power plant.

Net Power Output [MWe]	P_{max,sCO_2} [MPa]	TIT [°C]	$TES_{capacity}$ [hour]	SM [-]	DNI_{nom} [W/m^2]	$T_{amb,nom}$ [°C]
50	30	750	10	2.4	850	15

The thermo-economic performance of the reference power plant in Table 2 considering different cycle layouts is presented in Figure 2 for a pressure range deemed affordable in the medium term (15 to 40 MPa). Akin to the results provided in [8], η_{th} presents an increasing trend with pressure up 35 MPa. This is common to all cycles except the Recompression cycle (RC), in which the efficiency peaks a little sooner (30 MPa). On the other hand, the minimum Overnight Capital Cost is found between 30 and 35 MPa for all the cycles considered. Based on this, turbine inlet pressure is set to 30 MPa in the reference CSP plant, since this yields the best compromise between thermodynamic and economic performances and has already been proven to be achievable in similar power technologies such as ultra-supercritical steam turbines [20].

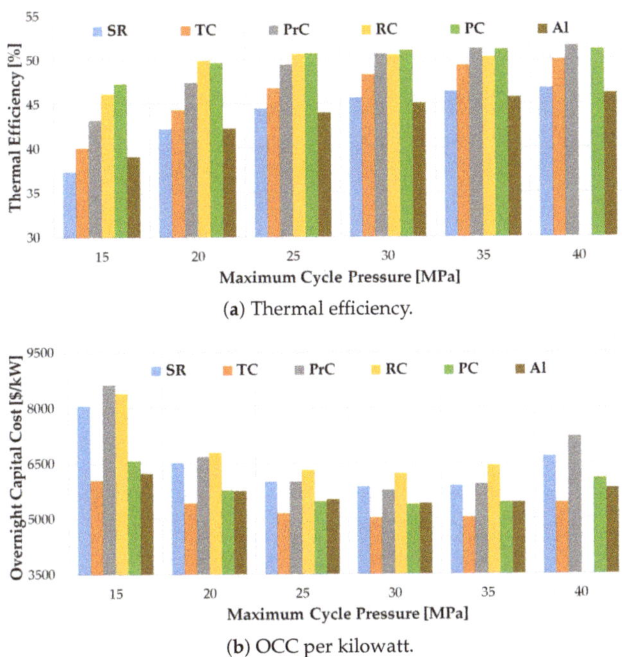

(a) Thermal efficiency.

(b) OCC per kilowatt.

Figure 2. Effect of maximum cycle pressure on thermal efficiency and Overnight Capital Cost per kilowatt. Legend: Simple Recuperated (SR), Transcritical (TC), Partial Cooling (PC), Recompression (R), Pre-compression (PrC) and Allam (Al). Data adapted from [10].

2.2. Economic Assessment

The calculations presented in Figure 2b above account for uncertainty regarding the capital costs of components. Using a Montecarlo analysis, each component cost is assigned a probability density function whereby costs are randomly taken from a given range of values with different probability: the cost of components with a lower maturity are allowed to change in a wider range whereas this range is much narrower for well established technologies. According to these calculations, uncertainty can potentially change the capital cost of the power block by plus/minus one-third of the cost, which is aligned with the information provided in the work by Weiland et al. for the National Energy Technology Laboratory and Sandia National Laboratories in the USA. In this latter work, the authors relied on data provided by Original Equipment Manufacturers to estimate the installation costs of small and large power plants operating on gaseous and solid fossil fuels [21]. The results, which also accounted for uncertainty of the input data provided by vendors, were similar to those reported in [10] both in terms of the total values and the variability brought about by uncertainty. As discussed by Carlson et al., this is very likely due to the lack of a well established market that prevents engineering costs from being charged on very few clients [22].

The data presented in Figure 2 in the previous section are expanded in Table 3, the economic values of which are calculated for the 85% confidence interval based on the probability density functions discussed in [10]. The table provides information about the main thermodynamic features—thermal efficiency (η_{th}) and temperature rise across the solar receiver (ΔT_{solar})—and about the Overnight Capital Cost of the plant and the contribution of each major equipment (TES, Solar Field, Tower/Receiver and Power Block). According to these data, the *Transcritical CO$_2$* (TC) and *Recompression* (RC) layouts yield the lowest and highest installation costs, respectively, 5656 and 6867 $/kW$_{inst}$, while the Partial Cooling (PC) cycle seems to provide the best compromise between thermal and economic features, closely followed by the Allam (Al) cycle.

Table 3. Thermo-economic assessment of different cycle layouts. Adapted from [10].

Cycle	η_{th} [%]	ΔT_{solar} [°C]	OCC [$/kW$_{inst}$]	C_{TES} [k$]	C_{SF} [k$]	$C_{T\&R}$ [k$]	C_{PB} [k$]
SR	45.8	290	6404	78,184	85,657	75,123	50,585
TC	48.3	290	5656	76,648	80,675	75,307	21,896
PrC	50.6	254	6515	80,145	76,373	79,697	58,835
RC	50.5	220	6867	91,640	76,547	90,000	52,498
PC	51.1	290	5907	70,230	75,603	70,945	50,568
Al	45.0	290	5943	79,403	87,074	75,778	26,227

The information shown so far confirms that the installation costs based on sCO$_2$ power cycles are comparable or even lower (for some layouts) than for steam turbines used in state-of-the-art CSP plants —5800 $/kW$_{inst}$ according to [23]. The large cost share of the Thermal Energy Storage system is also confirmed, which puts the temperature rise across the solar receiver stems forward as a critical factor involved in plant design, given its very strong impact on the inventory of salts that is needed to operate a plant of given output and storage capacity. Finally, heat exchangers stem as the most relevant individual component in the power block cost-wise, Figure 3, with a larger share than turbomachinery; this confirms earlier comments in this section.

A closer look into the operation of the cycles compared in Table 3 reveals that the *Transcritical CO$_2$* cycle requires the most restricting boundary conditions in order to achieve the lowest installation costs in the list. This is due to the need to enable condensation of the working fluid, which is only possible if the inlet temperature to the *Transcritical CO$_2$* pump is set to about 15 °C (strictly speaking, a temperature lower than the critical temperature of CO$_2$ would suffice to enable the implementation of a *Transcritical* cycle. Nevertheless, performance-wise, this layout only makes sense if the saturation temperature in the condenser is substantially lower than the critical temperature. This is why a value of 15 °C at the pump inlet is usually adopted). This translates into ambient temperatures in the order of 8 °C if an evaporative cooling tower is considered or less than 5 °C if air coolers are used, values that are only rarely found in the arid (even desertic) locations where CSP plants are typically found. The dismissal of the *Transcritical CO$_2$* layout for these reasons narrows the selection of cost-effective configurations to the Allam and Partial Cooling cycles, the layouts of which are presented in Figure 4. The main features of these cycles are:

- Allam cycle: it is an extremely high recuperative cycle, an evolution of a standard Brayton cycle incorporating a three-step compression process with two compressors and a pump separated by an intercooler and a condenser respectively. Originally proposed by Allam [24] for oxy-combustion applications, this layout has been adapted considering pure sCO_2 as working fluid [8].
- Partial Cooling cycles: this cycle derives directly from Angelino's work [6], and it is an evolution of a Recompression cycle with the addition of a cooler and a pre-compressor before the flow-split. The most interesting features of the Partial Cooling cycle are its high specific work [25] and a very low sensitivity of global efficiency to deviations of pressure ratio from the optimum value [26].

The installation costs reported in Table 3 are presented graphically in Figure 3, showing that the high temperature heat exchanger (heat adder or heater) is the key contributor to installation costs, accounting for approximately half of the total OCC of the power block. This is due to the very high operating pressure and temperature of the sCO_2-FLiNaK heat exchanger which, as a consequence, requires the utilisation of special alloys like *Inconel 617* [10].

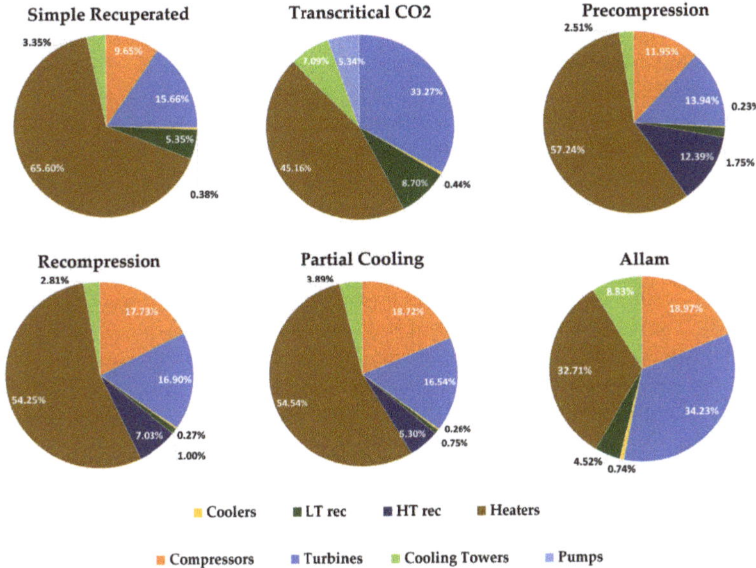

Figure 3. Breakdown of the Overnight Capital Cost of the power block.

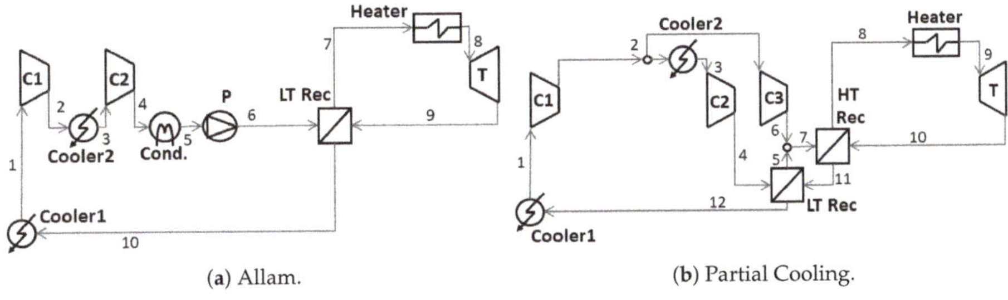

(a) Allam.

(b) Partial Cooling.

Figure 4. Layouts of the Allam and Partial Cooling cycles.

3. Levelised Cost of Electricity of CSP Plants Based on sCO$_2$

3.1. Preliminary Notes on the Assessment of the Levelised Cost of Electricity (LCoE)

In addition to the installation costs presented in earlier sections of this paper, estimating the Levelised Cost of Electricity of CSP plants based on sCO$_2$ technology requires a model to simulate the off-design performance of the power plant throughout the year. This provides the hourly output of the power plant for the specific boundary conditions—ambient temperature and pressure, available solar radiation, as described in [27] and, more recently [11]. The resulting information is then combined with the installation costs to obtain *LCoE*, as discussed in detail by M. Martin [13]. To this end, the authors have relied on the System Advisor Model SAM developed by the National Renewable Energy Laboratory of the United States' Department of Energy, since this is widely accepted by the scientific community and already employed in cost estimation for sCO$_2$ power plants [28,29].

The first part of this section describes the methodology used to incorporate the part-load performance of the Allam and Partial Cooling cycles into SAM's calculation procedure. Then, this information is used to model the reference 50 MW CSP plant in Table 2, considering either of the cycles selected in the previous section. For each one, two different locations and four different combinations of financial parameters and dispatch control models are assessed. Finally, the results obtained are compared against a state-of-the-art CSP plant using a standard power block based on steam turbines.

The input parameters defining the solar field are set to their default values, meaning that the geometry of the field is optimised according to state-of-the-art specifications that are representative of the current industrial practice. The type and composition of the molten salt (or Heat Transfer Fluid considered) is selected in the *Tower and Receiver* menu, where either commercial solar salts (NaNO$_3$KNO$_3$) or FLiNaK can be selected. The latter is the same molten salt employed by the in-house models developed by the authors and it is therefore selected for this *LCoE* assessment.

For the power block, SAM is limited to the *Recompression* layout when it comes to power blocks based on supercritical CO$_2$ systems, which is a strong limitation in this work. This is why the user-defined option is selected, which enables implementing the off-design performance of the Allam and Partial Cooling cycles modelled with the proprietary code developed by the authors and described in [11]. This is the step where the integration between SAM and the authors' in-house codes actually takes place, enabling a much more detailed and flexible assessment of supercritical CO$_2$ cycles than otherwise enabled by SAM. This integration requires the following specific information, in addition to more general technical features like the type of cooling system used (air or water cooled):

- MOD1-Performance as a function of HTF temperature: the part-load performance of the power cycle for variable molten salt (hot) temperature is obtained for three normalised mass flow rates of molten salts, in this case: 0.2, 1 and 1.05. The rated hot temperature of molten salts is set to 770 °C and this parameter is varied between 700 and 800 °C in the analysis.
- MOD2-Performance as a function of HTF mass flow rate: the part-load performance of the power cycles for variable mass flow rate of molten salts is obtained for three values of ambient temperature, in this case: 5, 15 and 40 °C. To this end, the same range of normalised mass flow rate as in the previous bullet point is considered; i.e., between 20% and 105% of the rated mass flow rate of FLiNaK .
- MOD3-Performance as a function of ambient temperature: the part-load performance of the power cycles is obtained when ambient temperature varies between 5 and 40 °C. Again, three molten salt (hot) temperatures are considered: 700, 770 and 800 °C.

The MOD-1 to MOD-3 tables corresponding to the reference power plant based on the Allam and Partial Cooling cycles are provided in Appendix A. They contain information about the specific off-design cycle performance considered in the simulations leading to the calculation of the Levelised Cost of Electricity and, therefore, they are deemed very valuable in terms of credibility of the work. The following list clarifies each parameter reported in Appendix A:

- Net electric output ($\dot{W}_{cycle}$): power output at generator terminals minus auxiliary power consumption. Auxiliary power accounts for the power needed to drive auxiliary equipment in the power block.
- Heat input (Q_{in}): heat supplied to the power block (i.e., absorbed by the working fluid).
- Power consumption of the cooling system ($\dot{W}_{cooling}$): power consumed by the evaporative cooling system.
- Water mass flow rate of the cooling system ($\dot{m}_{w,cooling}$): total amount of water needed in the cooling system.

These parameters are obtained for the so-called best control strategy and they must be normalised with respect to their rated values. For a more detailed discussion about how the power cycles are operated off-design, interested readers are referred to a complementary work by the authors [11].

3.2. Dispatch Control and Financial Model

Dispatch control refers to the scheme under which the Thermal Energy Storage System and power block are operated. In other words, the periods and operating conditions during which the power block drains energy from the Thermal Energy Storage to make up for the lack of concentrated solar energy supplied from the solar field directly (in a power plant using molten salts, heat input to the power block is always provided by the Thermal Energy Storage system. The text refers to situations where the net energy balance in the Thermal Energy Storage systems is negative), in order to achieve the desired turbine output.

Two different dispatch control schemes are considered in this work: the default scheme reported by SAM and the scheme proposed by the SunShot Vision Study [28]. A graphical representation of these is presented in Figure 5 which confirms that these schemes merely define an overall schedule of plant operation but they do not substitute or interfere with the off-design control strategies described in another work from the same authors [11]. Actually, the control schemes are aimed at enabling the output setting assigned by the dispatchability scheme. In particular, for SAM's default dispatch control, the plant runs at full capacity whenever possible, enabling a 5% over-charge during the central hours of the day in summer time, characterised by the highest solar irradiation. Running the plant in these conditions increases the average output of the plant but this is, potentially, at the cost of fewer operating hours in seasons with fewer sun hours. As opposed to this, the dispatch control scheme adopted by the SunShot Vision case drains energy from the Thermal Energy Storage system to sustain operation at 75% load during off-sun hours or during the day in seasons with lower solar radiation. This reduced load operation, marked in dark green (also number 2) in Figure 5, enables longer operating hours and reduces the number of shut-down/start-up manoeuvrers throughout the year.

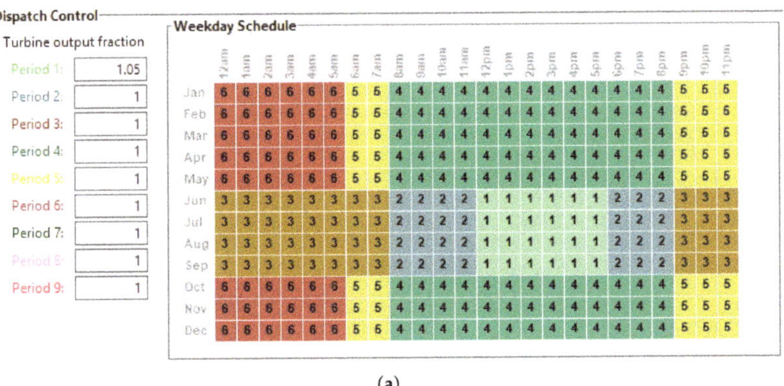

(a)

Figure 5. *Cont.*

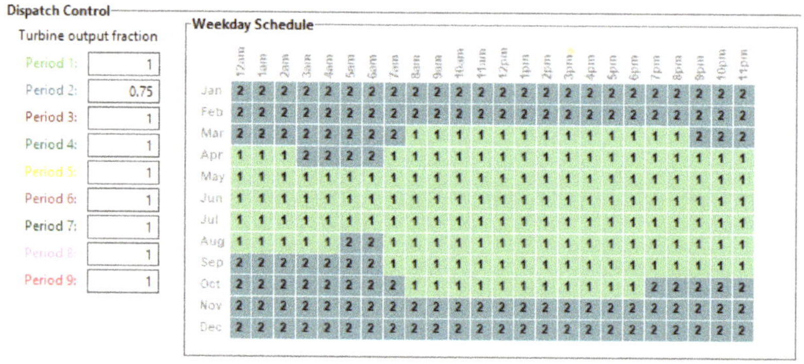

(b)

Figure 5. Different dispatch controls used to assess the Levelised Cost of Electricity. (**a**) System advisor model's (SAM's) default dispatch control. (**b**) Dispatch control scheme for the SunShot Vision case [28].

The different sets of financial assumptions adopted by SAM and the SunShot Vision study are listed in Table 4. The information includes the target internal rate of return and time (year) when this should be achieved, lifetime of the project, inflation rate, nominal discount rate (also termed rate of return), debt ratio, payback time and interest rate. These parameters are needed to calculate the corresponding *Power Purchase Agreement* (*PPA*) required to meet the target IRR according to the remaining financial settings. This *PPA* is actually another difference between the financial models based on SAM's and Sunshot cases, in particular regarding the *Time-of-delivery* factors (*TOD*) shown in Figure 6 and Table 5 which are multiplied by the *PPA* bid price to calculate the energy price in the corresponding hour. These *TODs* are therefore correction factors to take into account the variable sales price of electricity (*PPA*) throughout the day.

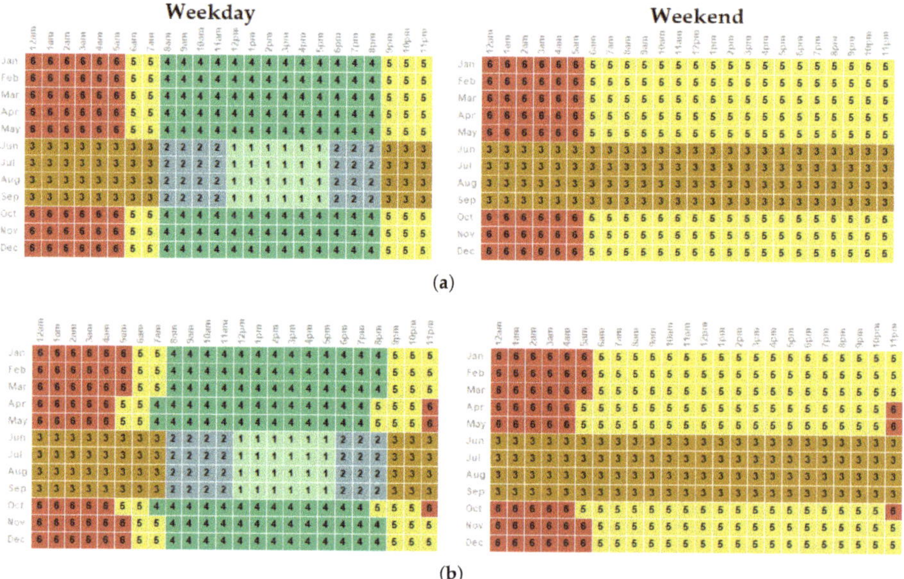

Figure 6. Different Time-of-Delivery factors used to assess the Levelised Cost of Electricity. (**a**) SAM's default Time-of-Delivery factors. (**b**) Delivery factors of the SunShot Vision Study case [28].

Table 4. Financial parameters employed in the SAM's default and SunShot Vision cases. The parameters that are not reported here are set to the values used by SAM's default model in Table 6.

	SAM Default	**SunShot Vision Study**
IRR target [%]	11	15
IRR target Year	20	30
Analysis period [years]	25	35
Inflation rate [%]	2.5	3
Nominal Discount rate [%]	8.14	8.66
Loan Percent of total capital cost [%]	50	60
Loan Duration [years]	18	15
Loan Annual interest rate [%]	7	7.1

Table 5. Different Time-of-Delivery factors employed by SAM's Default and SunShot Vision Study's financial models.

Financial Model	**Period 1**	**Period 2**	**Period 3**	**Period 4**	**Period 5**	**Period 6**
SAM Default	2.064	1.2	1	1.1	0.8	0.7
SunShot Vision Study	3.28	1.28	0.67	1.02	0.82	0.65

The combination of dispatchability and financial schemes described in this section yields the four cases listed in Table 6. These are studied in the next section to check the associated Levelised Cost of Electricity.

Table 6. Cases showing different combinations of Dispatch Control and Financial Models.

Case No.	**Financial Model**	**Dispatch Control**
Case 1	Default SAM	Default SAM
Case 2	Default SAM	SunShot Vision Study
Case 3	SunShot Vision Study	Default SAM
Case 4	SunShot Vision Study	SunShot Vision Study

3.3. Overall Analysis of the Levelised Cost of Electricity

The results obtained for the three CSP power plants described before, and for the four possible cases, are presented in Figures 7 and 8 for two reference locations in North America: Las Vegas and Tonopah (NV). These two locations are selected for their vicinity to existing Concentrated Solar Power plants using either collector technology, parabolic trough and central receiver, including the then first-of-a-kind Crescent Dunes project which made use of tower technology and an impressive Thermal Energy Storage system enabling operation at full capacity for ten hours. Five figures of merit are taken into account, spanning across the thermal and financial features of the plant:

- Thermal performance:

    - Yield (E_{year}), [GWh]: this is the annual production of electricity of the power plant.
    - Capacity Factor (CF), [%]: ratio from the system's annual production of electricity in the first year of operation to the theoretical energy production, should the system run at the rated capacity throughout the entire year. This is a measure of the electricity that the system would be able to produce if it were operated at its nominal capacity for every hour of the year, and it can be significantly affected by the plant location and by the operation (dispatch control) of the Thermal Energy Storage system.

- Financial:

    - Levelised Cost of Electricity $LCoE$ [¢/kWh]: a measure of the total project life cycle costs relative to the total production of energy throughout the entire project lifetime.

- Net Present Value *NPV* [$]: discounted (present) value of the net cash inflow.
- Internal Rate of Return *IRR*, [%]: the nominal discount rate that would yield null *NPV* for given economic and financial assumptions (including the sales price of electricity specified in the *Power Purchase Agreement—PPA*.

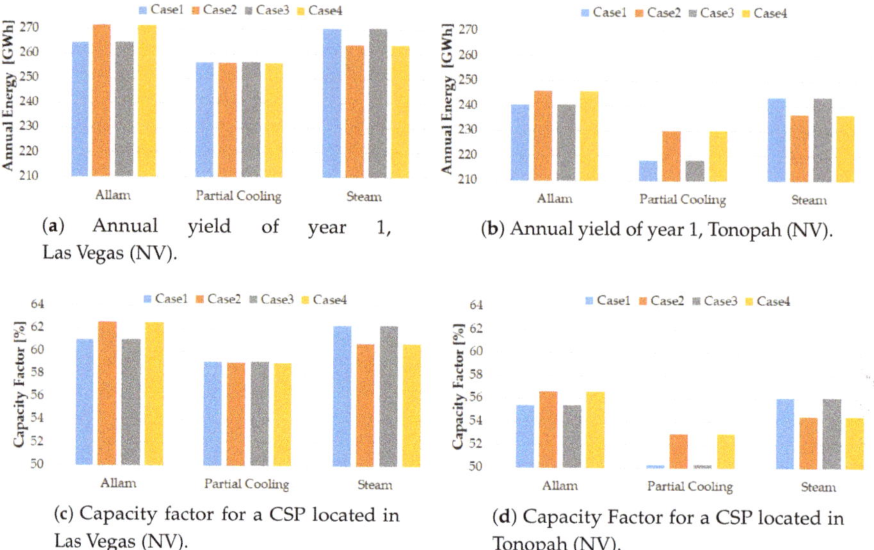

(**a**) Annual yield of year 1, Las Vegas (NV).

(**b**) Annual yield of year 1, Tonopah (NV).

(**c**) Capacity factor for a CSP located in Las Vegas (NV).

(**d**) Capacity Factor for a CSP located in Tonopah (NV).

Figure 7. Annual yield and capacity factor for the three power cycles and two locations considered. Designation of cases is explained in Table 6.

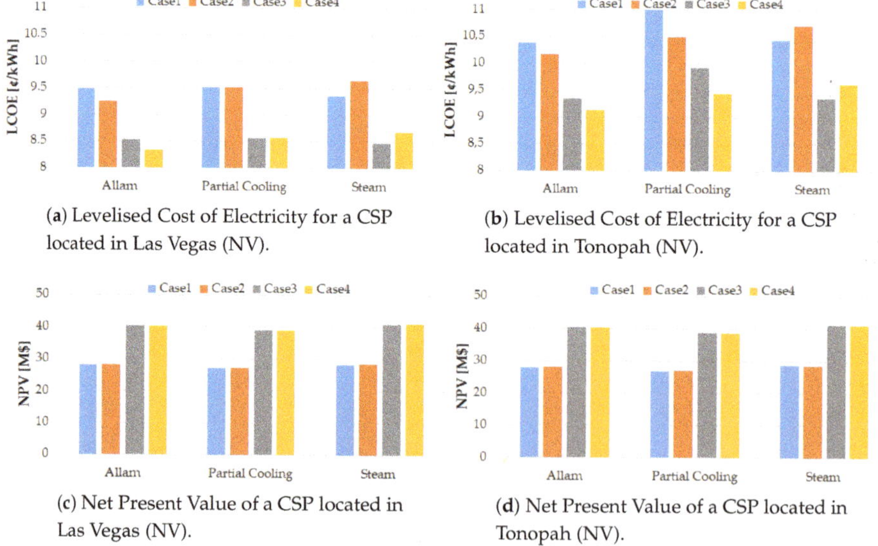

(**a**) Levelised Cost of Electricity for a CSP located in Las Vegas (NV).

(**b**) Levelised Cost of Electricity for a CSP located in Tonopah (NV).

(**c**) Net Present Value of a CSP located in Las Vegas (NV).

(**d**) Net Present Value of a CSP located in Tonopah (NV).

Figure 8. Levelised Cost of Electricity and Net Present Value of the three different cycles and four combinations of financial/dispatch control settings. Two locations in Nevada are considered: Las Vegas and Tonopah.

Several observations are worthwhile in Figure 7. The location of a CSP plant is expected to have a very strong impact on the annual yield. This is to be expected and is confirmed here by the larger production of electricity of the plant in Las Vegas.

Another interesting aspect, which is not as evident, is the impact of the dispatch control scheme, which yields a very variable plant performance pattern. For supercritical Carbon Dioxide plants based on the Allam cycle, the dispatch control proposed in SAM by default (Cases 1 and 3) is able to produce 5 GWh/year more energy than that proposed in the SunShot Vision Study. In order to find the reasons for this, it is reminded that the latter scheme was based on lower power settings to extend the operating time of the plant in periods with expectedly low solar availability, as opposed to the scheme proposed by SAM where power generation is maximised regardless of a potentially higher number of start-ups. The superior performance of this approach will have to be compared against the economic impact of the latter on maintenance costs in an actual power plant. Alas, as expected, the higher yield of Cases 1 and 3 in both locations translate into higher capacity factors.

Interestingly, the impact of dispatch control on plant performance for a conventional CSP plant based on a steam cycle is exactly the opposite. The dispatch control proposed in the SunShot Vision Study produces a higher yield than the default control proposed by SAM. The margin between the two is again in the order of 5 GWh/year, and the reasons for this are found in the off-design performance of the steam cycle, and to make it even more interesting, for a CSP plant using sCO_2 technology based on the Partial Cooling cycle, the annual production of electricity seems to be totally insensitive to the dispatchability scheme adopted, as shown in Figure 7. The patterns discussed are applicable to both locations, which supports their dependence on the characteristics of the power block and not on the boundary conditions of the power block. All this opens a very interesting research pathway incorporating the combined optimisation of both cycle technology, cycle layout and dispatch control scheme.

From a quantitative standpoint, the foregoing qualitative considerations translate as follows, for the cases considered. A CSP plant based on the Allam cycle in Las Vegas achieves 10% higher yield and capacity factor than if it were located in Tonopah and the difference would increase to about 12.5% if a Partial Cooling sCO_2 or a steam cycle were used. Regarding dispatch control, the SunShot Vision Study setting yields 3% higher E_{year} and CF when using the Allam cycle whereas a 3% drop in these parameters must be expected when considering a steam-turbine CSP plant. Further to the discussion in the previous paragraph, this can also be explained by the fact that SAM's Default and SunShot Vision Study's dispatch control modes are specifically designed for steam and sCO_2 power cycles respectively.

The same capacity to significantly change the results is not observed when assessing the two sets of financial assumptions. The input parameters considered in the SunShot Vision Study always lead to better financial results than SAM's, yielding higher NPV and lower $LCoE$ for a given location, as shown in Figure 8. This is mostly due to the longer lifetime and higher IRR considered, even if the former model presents a larger debt fraction, set to 60% (versus 50% for SAM's default case) of the total capital cost. It is also observed that NPV depends on the financial model given the minor deviations seen between different locations, cycles or dispatch control systems. On the other hand, $LCoE$ happens to be strongly affected by all the factors considered so far. In particular:

- Las Vegas yields lower $LCoE$, even if some $LCoEs$ obtained with the SunShot Vision Study case considering the Allam cycle in Tonopah are comparable to those obtained by the SAM setting in Las Vegas, regardless of the cycle used.
- The trend followed by $LCoE$ is approximately symmetrical to the figures of merit indexing thermal performance (CF and E_{year}) and balanced by the financial model. Higher CF usually comes with lower $LCoE$ but, if the two options with the lowest CF are considered—Partial Cooling cycle located in Tonopah for Cases 1 and 2—it is found that Case 1 always yields the highest $LCoE$ whereas the SunShot model can compensate for the CF effect in Case 3.

- If the same financial model is considered (Cases 1–2, Cases 3-4), the lowest *LCoEs* are achieved by plants presenting better performance metrics (higher *CF* and E_{year}), as observed in Figures 7a,b and 8a,b. Based on this foreseeable result, increasing the capacity factor of a plant (and therefore its annual yield) is confirmed to be of capital importance to increase the feasibility of sCO_2-based CSP plants.

Based on the foregoing results, it is difficult to ascertain the best power cycle, given that the three power cycles yield very similar *LCoE* in the order of 8.5–9.5 ¢/kWh, see Figure 8a,b. Moreover, the small differences observed lie within the cumulative uncertainty incurred by all the assumptions made throughout the analysis.

Figure 9 presents the final comparison of *LCoE* for the reference plant using the power cycles considered. The length of each bar comes determined by the values taken by *LCoE* for the different cases considered in Table 6. The lowest cost of electricity is obtained for the best combination of the Allam cycle (8.33 ¢/kWh), while the worst case based on the Partial Cooling cycle yields the highest *LCoE* (11.02 ¢/kWh). Interestingly, the range of *LCoE* for plants based on steam turbine technology falls entirely within the values covered by the other cycles. This confirms the lack of a clear, unambiguous conclusion about the optimum CSP plant concept stemming from this work, as suggested in the introductory section of the paper. To the authors' opinion, however, this result does confirm the large potential of sCO_2 power cycles for CSP applications, which must be considered as a solid alternative to the standard steam Rankine cycle approach in spite of the apparently marginal gain suggested by Figure 9. Such a strong statement is based on the fact that, even if the results obtained are still far from the 6 ¢/kWh target set by the SunShot programme (or 3 ¢/kWh in the longer term), it must not be forgotten that the *LCoE* values presented in this paper for sCO_2 are estimates based on deliberately conservative assumptions. For instance, the utilisation of molten salts that are less costly than FLiNaK, which is a reasonable assumption for an actual power plant in the future, would certainly cut down installation costs and, therefore, *LCoE* by a large fraction given the dominant role of the Thermal Energy Storage System in determining the economic performance of the plant. This is a very likely possibility in the near future which would cut LCoE down for the sCO_2 cases in Figure 9 but would not affect the cost of the steam-based case. Therefore, it may as well be the case that mid-term CSP plants employing Allam and Partial Cooling cycles are able to achieve *LCoEs* lower than those obtained in this research, and closer to the SunShot objective, once the very strong economies of scale that are characteristic of CSP start impacting sCO_2 power blocks. There is no reason why a cost decline similar to that experienced by conventional CSP-STE plants cannot be experienced by this new technology in the near term.

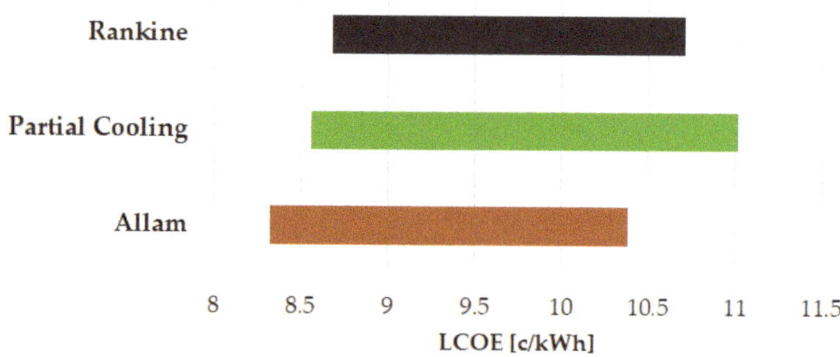

Figure 9. Range of Levelised Cost of Energy (*LCoE*) for CSP plants based on different power cycles.

With this in mind, it is concluded that more accurate part-load models and integration schemes in SAM or equivalent software will prove that either the Allam or Partial Cooling cycle layouts have the potential to make a strong case for the next generation of CSP plants based on sCO_2 power cycles, enabling the ambitious objectives targeted by the SunShot Program. In this regard, it is interesting to see how the selection of a particular set of assumptions (both financial, economic and thermal) has the potential to turn the *Recompression* cycle into a competitive option to achieve the 6 ¢/kWh target of the SunShot programme [28], which is certainly in contrast with the conclusions obtained in this work. Far from discussing the credibility of other research works, the latter statement aims to highlight that much more work on the assessment of appropriate sets of non-technical boundary conditions is still required.

4. Some Considerations about Uncertainty

Assessing a trustworthy and thorough feasibility analysis of CSP plants employing sCO_2 power blocks is very challenging, mostly due to the low TRL and MRL of sCO_2 technology and to the scarcity of reliable cost-related information. To overcome this limitation, this work is based on a series of assumptions, in order to reduce the complexity of the problem down to a manageable level. In this section, these assumptions are revisited with the aim to assess the reliability of the results, highlighting both the positive features and also the main flaws.

There are two types of uncertainty when exploring the potential of disruptive technologies: avoidable and unavoidable. Avoidable uncertainty that is incurred in the simulation of processes due to the lack of precise information about the operating conditions and specifications of certain components. The same applies to the calculation of capital and operating costs of a technology that is not commercially available yet. As far as this work is concerned, previous publications by the authors have justified the simplifying assumptions made to carry out the thermal assessment [8,11] and to estimate the capital costs of sCO_2 technology [9,10]. In this former case, the performance results have been validated against results in literature and also against experimental data [30]. For capital costs, given the lack of mass production of the equipment needed (sCO_2 compressors, expanders and heat exchangers and also high pressure and temperature solar receivers), a dedicated Monte Carlo- based uncertainty quantification analysis is incorporated in the calculations. In either case, uncertainty can be classified as avoidable inasmuch as the continuous development of the technology and, later, the deployment of a commercial plant will expectedly yield more accurate technical and economic information.

The financial boundary conditions set to calculate the Levelised Cost of Electricity of the technology is affected by the so-called unavoidable uncertainty. Future economic scenarios are not foreseeable and, more often than not, previsions are drastically altered by unpredictable political and/or social changes; the twenty-first century is still in its infancy and it has already seen several of these crises. As a consequence of this, setting the discount rate turns out an uncertain process that includes thorough considerations about how investments in a particular market might unfold in the future. The same applies to other local assumptions related to foreseen energy policies and subsidies, especially in unstable regions of the world [31]. These aspects involved in the economic valuation of constructing innovative CSP plants are inherently uncertain and, therefore, cannot be removed from the analysis completely.

With all this in mind, the results obtained are deemed as trustworthy as possible for a technology that is still far from the marketplace. Moreover, as reported in the concluding remark of the previous section, the results obtained are thought to provide a very solid upper limit of the expected *LCoE* enabled by sCO_2 power cycles in CSP applications. From here, future works will continuously reduce avoidable uncertainty to values comparable to those reported for state-of-the-art CSP plants based on steam turbines.

5. Conclusions

This paper is focused on the assessment of the Levelised Costs of Electricity that should be expected if supercritical CO_2 power cycles were adopted in next generation Concentrated Solar Power plants for Solar Thermal Electricity generation. A simulation platform developed by the authors for the techno-economic analysis of Concentrated Solar Power plants has been integrated with the

Solar Advisor Model to produce a larger simulation tool enabling the assessment of technical and non-technical plant characteristics to a level of detail not enabled by the latter software alone.

Plants based on steam turbine technology and on sCO_2 cycles have been analysed under different financial/economic boundary conditions and for two different locations which can be regarded as above-average for CSP installations. In all cases, the net output of the plant is 50 MW and the Thermal Energy Storage system is sized to enable operation at full capacity for ten hours. In the most favourable cases, the Partial Cooling and Allam cycles provide an *LCoE* of 8.56 ¢/kWh and 8.33 ¢/kWh respectively, along with a Capacity Factor of around 59% and 62.5%. On the contrary, in the worst case, the *LCoE* of the Partial Cooling cycle increases to 11 ¢/kWh, with a *CF* slightly higher than 50%, while the values of the Allam cycle are 10.38 ¢/kWh and 55.4% respectively.

When these figures of merit are compared with a reference plant based on a state-of-the-art steam Rankine cycle, the main finding is that both sCO_2 cycles have the potential to yield *LCoEs* comparable to those of the reference plant, or even lower. Taking into account the conservative approach employed throughout the present work, especially in terms of installation costs, this comes to confirm that sCO_2 power cycles are an interesting alternative to enhance the competitiveness of CSP-STE plants in the mid to long term, even if the drastic cost reductions claimed by some authors seem not to be so straightforward in the short term. In the longer term, mass deployment and a further refinement of the technology (technology/cost-wise) will very likely increase the economic gains of CSP-sCO_2 plants but, at the moment, this remains yet to be verified.

In this regard, it is worth highlighting the groundbreaking concept proposed by the SCARABEUS project, funded by the European Commission and running from 2019 to 2023. In this project, CO_2 is doped with certain compounds to modify the critical properties of the resulting mixture, shifting the critical temperature to higher values and enabling the practical implementation of condensing sCO_2 cycles (for instance, the *Transcritical CO_2* cycle would be realisable even at very high ambient temperatures). Substituting these new CO_2 mixtures for pure CO_2 in the same layouts would, in turn, boost the efficiency of the resulting power plant, which translates into a smaller footprint of the solar field and also smaller size of the Thermal Energy Storage system. If these were eventually possible, as already suggested by the preliminary results in [3], the resulting power plant would easily reduce the LCoE reported in this paper by a large margin, possibly achieving the ambitious targets set forth by the SunShot programme.

Author Contributions: Conceptualization, D.S. and T. S.-L.; formal analysis, G.S.M.; funding acquisition, D.S.; investigation, F.C.; methodology, F.C., D.S. and F.J. J.-E.; software, F.C., G.S.M. and F.J.J.-E.; supervision, T.S.-L.; validation, T.S.-L. and F.J.J.-E.; writing—original draft, F.C.; writing—review and editing, D.S. and T.S.-L. All authors have read and agreed to the published version of the manuscript.

Funding: The SCARABEUS project has received funding from the European Union's Horizon 2020 research and innovation programme under grant agreement N ° 814985.

Acknowledgments: SoftInWay is gratefully acknowledged for supporting the simulation of turbomachinery performance maps for supercritical CO_2 applications.

Conflicts of Interest: The authors declare no conflict of interest.

Abbreviations

The following abbreviations are used in this manuscript:

LCoE	Levelised Cost of Energy
CSP	Concentrated Solar Power
PV	Photovoltaic
sCO_2	Supercritical CO_2
USD	US Dollar
TES	Thermal Energy Storage
HTF	Heat Transfer Fluid
TIT	Turbine Inlet Temperature

SR	Simple Recuperated cycle
TC	Transcritical CO2 cycle
PrC	Precompression cycle
RC	Recompression CO2 cycle
Al	Allam cycle
PC	Partial Cooling cycle
OCC	Overnight Capital Cost
SM	Solar Multiple
DNI	Direct Normal Irradiance
ΔT_{solar}	Temperature rise across solar receiver
SF	Solar Field
T&R	Tower and Receiver
PB	Power Block
LT Rec	Low-temperature receiver
HT Rec	High-temperature receiver
RMB	Chinese Renminbi
SAM	System Advisor model
$\dot{W}_{cycle}$	Power Cycle Net Electric Output
Q_{in}	Heat Input to Power Cycle
$\dot{W}_{cooling}$	Cooling System Power Consumption
$\dot{m}_{w,cooling}$	Cooling System Water Mass Flow Rate

Appendix A. Integrating the Off-Design Performance of the Power Block into the System Advisor Model

Table A1. Performance as a function of HTF temperature (MOD1) table for Allam cycle. "Low", "On" and "High" respectively refer to the three normalised mass flow employed: 0.2, 1 and 1.05.

T_{HTF}	$\dot{W}_{cycle,low}$	$\dot{W}_{cycle,on}$	$\dot{W}_{cycle,high}$	$Q_{in,low}$	$Q_{in,on}$	$Q_{in,high}$	$\dot{W}_{cooling,low}$	$\dot{W}_{cooling,on}$	$\dot{W}_{cooling,high}$	$\dot{m}_{w,cooling,low}$	$\dot{m}_{w,cooling,on}$	$\dot{m}_{w,cooling,high}$
700	0.13153	0.71357	0.77024	0.15189	0.75096	0.80142	0.051941	0.78613	0.93529	0.15189	0.75096	0.80142
702.5	0.13394	0.74109	0.78122	0.15365	0.76956	0.81041	0.053356	0.91307	0.9481	0.15365	0.76956	0.81041
705	0.13729	0.75097	0.78691	0.15541	0.77679	0.81566	0.056937	0.93796	0.94818	0.15541	0.77679	0.81566
707.5	0.13933	0.76061	0.79881	0.15708	0.78553	0.82476	0.057959	0.94214	0.95883	0.15708	0.78553	0.82476
710	0.14114	0.77012	0.8075	0.15887	0.7942	0.83404	0.057959	0.94521	0.95972	0.15887	0.7942	0.83404
712.5	0.1452	0.77963	0.81809	0.1606	0.80278	0.84288	0.06362	0.95214	0.96625	0.1606	0.80278	0.84288
715	0.14687	0.78942	0.82727	0.1623	0.81163	0.85203	0.06362	0.95657	0.96844	0.1623	0.81163	0.85203
717.5	0.14862	0.79871	0.83717	0.164	0.82002	0.86097	0.06362	0.96154	0.97291	0.164	0.82002	0.86097
720	0.1532	0.80834	0.84642	0.16577	0.82868	0.87006	0.07148	0.96511	0.97536	0.16577	0.82868	0.87006
722.5	0.15495	0.81803	0.85762	0.16746	0.83737	0.87924	0.07182	0.96885	0.97736	0.16746	0.83737	0.87924
725	0.1567	0.82762	0.86862	0.16917	0.84605	0.88846	0.07182	0.97157	0.97983	0.16917	0.84605	0.88846
727.5	0.1587	0.83734	0.87799	0.17078	0.85477	0.89726	0.072099	0.97444	0.98243	0.17078	0.85477	0.89726
730	0.16376	0.84674	0.88763	0.17264	0.8634	0.90657	0.084065	0.97719	0.98462	0.17264	0.8634	0.90657
732.5	0.16541	0.85598	0.897	0.17436	0.87206	0.91542	0.084065	0.97948	0.9869	0.17436	0.87206	0.91542
735	0.16721	0.86541	0.90685	0.17612	0.8807	0.92472	0.084065	0.98199	0.98892	0.17612	0.8807	0.92472
737.5	0.17044	0.87466	0.91619	0.17786	0.88933	0.93353	0.086255	0.98376	0.99081	0.17786	0.88933	0.93353
740	0.17475	0.88398	0.92624	0.17954	0.89795	0.94289	0.099483	0.98644	0.99231	0.17954	0.89795	0.94289
742.5	0.17609	0.8926	0.93621	0.18126	0.90633	0.95194	0.099483	0.98747	0.99356	0.18126	0.90633	0.95194
745	0.17818	0.9037	0.94623	0.18303	0.91525	0.96096	0.099483	0.98938	0.99468	0.18303	0.91525	0.96096
747.5	0.18009	0.9129	0.95629	0.18477	0.92387	0.97	0.099483	0.99106	0.99595	0.18477	0.92387	0.97
750	0.18185	0.92217	0.96617	0.18648	0.93249	0.97881	0.099483	0.99292	0.99647	0.18648	0.93249	0.97881
752.5	0.18395	0.93131	0.97666	0.18819	0.94105	0.98812	0.22354	0.99444	0.99689	0.18819	0.94105	0.98812
755	0.18638	0.94067	0.98576	0.18993	0.94962	0.99714	0.25828	0.99599	0.99718	0.18993	0.94962	0.99714
757.5	0.18853	0.9503	0.98962	0.19166	0.95825	1.0064	0.26095	0.9973	1.0055	0.19166	0.95825	1.0064
760	0.19122	0.96005	1.0051	0.19338	0.96687	1.015	0.26826	0.99824	1.0108	0.19338	0.96687	1.015
762.5	0.19333	0.96973	1.0153	0.19509	0.97547	1.0242	0.26849	0.99911	1.0135	0.19509	0.97547	1.0242
765	0.19544	0.97942	1.0254	0.19682	0.98412	1.0333	0.26861	0.99979	1.0156	0.19682	0.98412	1.0333
767.5	0.19694	0.98921	1.0356	0.19775	0.99273	1.0423	0.26884	1.0002	1.0177	0.19775	0.99273	1.0423
770	0.19793	1	1.0457	0.19837	1	1.0514	0.26884	1	1.0199	0.19837	1	1.0514
772.5	0.20412	1.0072	1.0559	0.20198	1.0098	1.0605	0.28955	1.0085	1.022	0.20198	1.0098	1.0605
775	0.20611	1.0174	1.0661	0.20374	1.0188	1.0695	0.2897	1.0111	1.024	0.20374	1.0188	1.0695
777.5	0.20803	1.0268	1.0763	0.20546	1.0273	1.0786	0.28983	1.0135	1.0259	0.20546	1.0273	1.0786
780	0.20896	1.0369	1.0865	0.20633	1.0361	1.0876	0.28968	1.0154	1.0278	0.20633	1.0361	1.0876
782.5	0.21399	1.0462	1.0968	0.20895	1.0445	1.0967	0.29744	1.0174	1.0295	0.20895	1.0445	1.0967
785	0.21598	1.0564	1.107	0.21066	1.0534	1.1057	0.29761	1.0186	1.0312	0.21066	1.0534	1.1057
787.5	0.21795	1.0657	1.1173	0.21235	1.0618	1.1148	0.30158	1.0198	1.0322	0.21235	1.0618	1.1148
790	0.22012	1.0759	1.1275	0.21412	1.0706	1.1239	0.30168	1.0208	1.0327	0.21412	1.0706	1.1239
792.5	0.2219	1.0851	1.1378	0.21584	1.0789	1.133	0.30599	1.0218	1.0331	0.21584	1.0789	1.133
795	0.22404	1.0952	1.1483	0.21756	1.0877	1.1421	0.30633	1.0229	1.0336	0.21756	1.0877	1.1421
797.5	0.22617	1.1033	1.1567	0.21928	1.0925	1.147	0.30664	1.0221	1.0327	0.21928	1.0925	1.147
800	0.22828	1.1074	1.1611	0.22099	1.0947	1.1493	0.30696	1.0221	1.0328	0.22099	1.0947	1.1493

Table A2. Performance as a function of HTF mass flow rate (MOD2) table for Allam cycle. "Low", "On" and "High" respectively refer to the three values of ambient temperature employed: 5, 15 and 40 °C.

T_{HTF}	$\dot{W}_{cycle,low}$	$\dot{W}_{cycle,on}$	$\dot{W}_{cycle,high}$	$Q_{in,low}$	$Q_{in,on}$	$Q_{in,high}$	$\dot{W}_{cooling,low}$	$\dot{W}_{cooling,on}$	$\dot{W}_{cooling,high}$	$\dot{m}_{w,cooling,low}$	$\dot{m}_{w,cooling on}$	$\dot{m}_{w,cooling,high}$
0.2	0.1069	-0.039852	-0.12903	0.11701	0.068519	0.068519	0.053491	0.15938	0.35023	0.11701	0.068519	0.068519
0.21	0.12236	-0.019735	-0.10588	0.13174	0.084742	0.084742	0.053491	0.15938	0.35023	0.13174	0.084742	0.084742
0.22	0.1377	0.00020115	-0.082962	0.14637	0.10085	0.10085	0.053491	0.15938	0.35023	0.14637	0.10085	0.10085
0.23	0.15294	0.019956	-0.060273	0.16091	0.11684	0.11684	0.053491	0.15938	0.35023	0.16091	0.11684	0.11684
0.24	0.16807	0.039529	-0.037816	0.17536	0.13271	0.13271	0.053491	0.15938	0.35023	0.17536	0.13271	0.13271
0.25	0.18309	0.05892	-0.015588	0.18971	0.14847	0.14847	0.053491	0.15938	0.35023	0.18971	0.14847	0.14847
0.26	0.198	0.078131	0.0064081	0.20397	0.16411	0.16411	0.053491	0.15938	0.35023	0.20397	0.16411	0.16411
0.27	0.2128	0.097159	0.028174	0.21813	0.17964	0.17964	0.053491	0.15938	0.35023	0.21813	0.17964	0.17964
0.28	0.22749	0.11601	0.049709	0.23221	0.19505	0.19505	0.053491	0.15938	0.35023	0.23221	0.19505	0.19505
0.29	0.24207	0.13467	0.071013	0.24619	0.21034	0.21034	0.053491	0.15938	0.35023	0.24619	0.21034	0.21034
0.3	0.25655	0.15316	0.092087	0.26007	0.22552	0.22552	0.053491	0.15938	0.35023	0.26007	0.22552	0.22552
0.31	0.27091	0.17146	0.11293	0.27387	0.24057	0.24057	0.053491	0.15938	0.35023	0.27387	0.24057	0.24057
0.32	0.28517	0.18958	0.13354	0.28756	0.25552	0.25552	0.053491	0.15938	0.35023	0.28756	0.25552	0.25552
0.33	0.29931	0.20752	0.15392	0.30117	0.27034	0.27034	0.053491	0.15938	0.35023	0.30117	0.27034	0.27034
0.34	0.31335	0.22528	0.17407	0.31468	0.28505	0.28505	0.053491	0.15938	0.35023	0.31468	0.28505	0.28505
0.35	0.32728	0.24286	0.19399	0.3281	0.29965	0.29965	0.053491	0.15938	0.35023	0.3281	0.29965	0.29965
0.36	0.34109	0.26025	0.21368	0.34143	0.31413	0.31413	0.053491	0.15938	0.35023	0.34143	0.31413	0.31413
0.37	0.3548	0.27747	0.23314	0.35466	0.32849	0.32849	0.053491	0.15938	0.35023	0.35466	0.32849	0.32849
0.38	0.3684	0.2945	0.25237	0.3678	0.34273	0.34273	0.053491	0.15938	0.35023	0.3678	0.34273	0.34273
0.39	0.38189	0.31135	0.22856	0.38085	0.35686	0.32091	0.053491	0.15938	0.35023	0.38085	0.35686	0.32091
0.4	0.39527	0.32802	0.26085	0.3938	0.37087	0.34112	0.053491	0.15938	0.36782	0.3938	0.37087	0.34112
0.41	0.40854	0.34451	0.28613	0.40666	0.38477	0.35995	0.053491	0.15938	0.38728	0.40666	0.38477	0.35995
0.42	0.42171	0.36082	0.3164	0.41943	0.39855	0.37812	0.053491	0.15938	0.40453	0.41943	0.39855	0.37812
0.43	0.43476	0.37694	0.33605	0.4321	0.41221	0.39545	0.053491	0.15938	0.42904	0.4321	0.41221	0.39545
0.44	0.4477	0.39289	0.36078	0.44468	0.42575	0.41214	0.053491	0.15938	0.44911	0.44468	0.42575	0.41214
0.45	0.46054	0.38321	0.38195	0.45717	0.4292	0.42861	0.053491	0.15938	0.47182	0.45717	0.4292	0.42861
0.46	0.47326	0.404	0.4102	0.46956	0.44513	0.44474	0.053491	0.16552	0.47382	0.46956	0.44513	0.44474
0.47	0.48588	0.43166	0.41673	0.48186	0.46096	0.45969	0.053491	0.18883	0.47382	0.48186	0.46096	0.45969
0.48	0.49838	0.44323	0.43972	0.49407	0.47608	0.47291	0.053491	0.19835	0.93402	0.49407	0.47608	0.47291
0.49	0.51078	0.46693	0.45956	0.50618	0.49131	0.48742	0.053491	0.23203	0.98613	0.50618	0.49131	0.48742
0.5	0.52307	0.48585	0.48575	0.5182	0.50617	0.502	0.053491	0.24617	1.0357	0.5182	0.50617	0.502
0.51	0.53525	0.50719	0.49279	0.53013	0.52041	0.51576	0.053491	0.26603	1.0791	0.53013	0.52041	0.51576
0.52	0.54732	0.52366	0.51066	0.54196	0.53333	0.52972	0.053491	0.26942	1.0791	0.54196	0.53333	0.52972
0.53	0.55928	0.53259	0.53142	0.5537	0.54574	0.54392	0.053491	0.26942	1.1096	0.5537	0.54574	0.54392
0.54	0.57113	0.5484	0.54409	0.56535	0.55973	0.55667	0.053491	0.26942	1.1184	0.56535	0.55973	0.55667
0.55	0.58287	0.56828	0.55687	0.5769	0.57314	0.56974	0.053491	0.37168	1.1187	0.5769	0.57314	0.56974
0.56	0.58357	0.57193	0.5668	0.58237	0.57674	0.57973	0.053491	0.72927	1.1187	0.58237	0.57674	0.57973
0.57	0.59834	0.58589	0.58081	0.59522	0.58986	0.58715	0.067841	0.80783	1.1441	0.59522	0.58986	0.58715
0.58	0.60928	0.59789	0.59414	0.6069	0.60168	0.59923	0.069021	0.81207	1.1474	0.6069	0.60168	0.59923
0.59	0.62779	0.61627	0.61141	0.62199	0.61559	0.61279	0.13677	0.85494	1.1623	0.62199	0.61559	0.61279
0.6	0.64146	0.62927	0.62563	0.63429	0.62764	0.62478	0.14465	0.86535	1.1694	0.63429	0.62764	0.62478
0.61	0.64951	0.64051	0.63781	0.6435	0.63844	0.63575	0.52769	0.86614	1.1699	0.6435	0.63844	0.63575
0.62	0.66367	0.65191	0.64891	0.65589	0.64962	0.64648	0.59751	0.86716	1.1714	0.65589	0.64962	0.64648
0.63	0.67562	0.67302	0.6702	0.66758	0.66541	0.66283	0.61225	0.90866	1.196	0.66758	0.66541	0.66283
0.64	0.68855	0.68398	0.68196	0.67948	0.67656	0.67385	0.63344	0.91034	1.1982	0.67948	0.67656	0.67385
0.65	0.70059	0.69831	0.69662	0.69092	0.68881	0.68651	0.64752	0.9243	1.2083	0.69092	0.68881	0.68651
0.66	0.71025	0.71011	0.71094	0.70105	0.69996	0.69764	0.6481	0.93036	1.2103	0.70105	0.69996	0.69764
0.67	0.72009	0.71983	0.72496	0.71116	0.70997	0.70915	0.64864	0.93094	1.2161	0.71116	0.70997	0.70915
0.68	0.72955	0.7344	0.72733	0.72101	0.72036	0.71783	0.64911	0.93125	1.2174	0.72101	0.72036	0.71783
0.69	0.74311	0.73725	0.73662	0.73118	0.72923	0.72749	0.64935	0.9323	1.2181	0.73118	0.72923	0.72749
0.7	0.74831	0.74647	0.74576	0.74015	0.73877	0.73693	0.66244	0.93287	1.2188	0.74015	0.73877	0.73693
0.71	0.75862	0.75682	0.75605	0.75041	0.74839	0.74614	0.67003	0.948	1.233	0.75041	0.74839	0.74614
0.72	0.76841	0.76737	0.76742	0.76036	0.75866	0.75673	0.67626	0.95423	1.2385	0.76036	0.75866	0.75673
0.73	0.77798	0.77702	0.77572	0.77021	0.76845	0.76663	0.68167	0.95672	1.2414	0.77021	0.76845	0.76663
0.74	0.78731	0.78746	0.78675	0.77992	0.77845	0.77656	0.68586	0.96094	1.244	0.77992	0.77845	0.77656
0.75	0.79698	0.79546	0.79737	0.78943	0.78791	0.78639	0.68972	0.96401	1.2462	0.78943	0.78791	0.78639
0.76	0.80592	0.80577	0.80778	0.79892	0.79742	0.79606	0.69329	0.96638	1.2485	0.79892	0.79742	0.79606
0.77	0.81464	0.81603	0.81758	0.80801	0.80685	0.80563	0.6959	0.96847	1.2512	0.80801	0.80685	0.80563
0.78	0.82487	0.82602	0.82407	0.81743	0.81635	0.81463	0.69824	0.97078	1.2542	0.81743	0.81635	0.81463
0.79	0.83453	0.83418	0.83121	0.82661	0.82532	0.82355	0.7006	0.97303	1.2569	0.82661	0.82532	0.82355
0.8	0.84263	0.83954	0.83928	0.83551	0.83402	0.83256	0.70294	0.9763	1.2586	0.83551	0.83402	0.83256
0.81	0.84783	0.84739	0.84912	0.84391	0.84281	0.84171	0.70601	0.9781	1.2606	0.84391	0.84281	0.84171
0.82	0.85559	0.85693	0.85753	0.85262	0.8517	0.85045	0.70796	0.97976	1.2622	0.85262	0.8517	0.85045
0.83	0.86501	0.86518	0.86571	0.86147	0.86036	0.8592	0.70945	0.98146	1.264	0.86147	0.86036	0.8592
0.84	0.87321	0.87341	0.87364	0.87013	0.86905	0.8679	0.71107	0.98295	1.2659	0.87013	0.86905	0.8679
0.85	0.88117	0.88117	0.88155	0.87858	0.87756	0.87647	0.71257	0.98467	1.2677	0.87858	0.87756	0.87647
0.86	0.88893	0.88902	0.88927	0.88704	0.88605	0.88489	0.71404	0.98617	1.2691	0.88704	0.88605	0.88489
0.87	0.89644	0.89673	0.89833	0.89538	0.89444	0.89344	0.71548	0.98759	1.2704	0.89538	0.89444	0.89344
0.88	0.90459	0.90568	0.90648	0.90385	0.90303	0.90192	0.71669	0.98876	1.2718	0.90385	0.90303	0.90192
0.89	0.91306	0.91369	0.91448	0.91215	0.91132	0.91023	0.71771	0.99001	1.2732	0.91215	0.91132	0.91023
0.9	0.92105	0.92173	0.9233	0.92044	0.91967	0.91865	0.71886	0.99112	1.2744	0.92044	0.91967	0.91865
0.91	0.9291	0.93038	0.92977	0.92875	0.92784	0.92679	0.71988	0.9921	1.2762	0.92875	0.92784	0.92679
0.92	0.93709	0.93654	0.93789	0.9368	0.93581	0.93504	0.72072	0.99371	1.2774	0.9368	0.93581	0.93504
0.93	0.94381	0.94497	0.9462	0.94485	0.94411	0.94319	0.72162	0.99468	1.2785	0.94485	0.94411	0.94319
0.94	0.95238	0.9531	0.95408	0.95298	0.9522	0.95122	0.7222	0.99573	1.2797	0.95298	0.9522	0.95122
0.95	0.96021	0.96099	0.96201	0.96096	0.96021	0.95931	0.72281	0.99665	1.2809	0.96096	0.96021	0.95931
0.96	0.96815	0.96896	0.97002	0.96904	0.96829	0.96743	0.72344	0.99767	1.282	0.96904	0.96829	0.96743
0.97	0.97583	0.97668	0.97755	0.97695	0.97625	0.97537	0.72408	0.99855	1.2831	0.97695	0.97625	0.97537
0.98	0.98335	0.98478	0.98585	0.98485	0.98422	0.98346	0.72465	0.99949	1.2844	0.98485	0.98422	0.98346
0.99	0.9914	0.99224	0.99348	0.99266	0.9921	0.9914	0.72522	0.99977	1.2853	0.99266	0.9921	0.9914
1	0.99916	1	1.0012	1.0006	1	0.99932	0.72584	1	1.2866	1.0006	1	0.99932
1.01	1.007	1.0078	1.0091	1.0085	1.0079	1.0072	0.72635	1.0002	1.2874	1.0085	1.0079	1.0072
1.02	1.0148	1.0156	1.0168	1.0162	1.0155	1.0148	0.72689	1.0005	1.2877	1.0162	1.0155	1.0148
1.03	1.0229	1.0236	1.0248	1.0241	1.0234	1.0227	0.7275	1.0008	1.288	1.0241	1.0234	1.0227
1.04	1.0307	1.0314	1.0325	1.032	1.0313	1.0307	0.7281	1.0011	1.2885	1.032	1.0313	1.0307
1.05	1.0354	1.0375	1.0376	1.0385	1.0373	1.0362	0.72903	0.99815	1.285	1.0385	1.0373	1.0362

Table A3. Performance as a function of ambient temperature (MOD3) table for Allam cycle. "Low", "On" and "High" respectively refer to the three values of molten salt (hot) temperature levels employed: 700, 770 and 800 °C.

T_{HTF}	$\dot{W}_{cycle,low}$	$\dot{W}_{cycle,on}$	$\dot{W}_{cycle,high}$	$Q_{in,low}$	$Q_{in,on}$	$Q_{in,high}$	$\dot{W}_{cooling,low}$	$\dot{W}_{cooling,on}$	$\dot{W}_{cooling,high}$	$\dot{m}_{w,cooling,low}$	$\dot{m}_{w,cooling on}$	$\dot{m}_{w,cooling,high}$
5	0.84245	1.0002	1.0628	0.88637	1.0012	1.043	0.68615	0.74597	0.74426	0.88637	1.0012	1.043
6.25	0.83983	1.0029	1.0651	0.88229	1.001	1.0423	0.69913	0.77435	0.77091	0.88229	1.001	1.0423
7.5	0.83769	0.9998	1.0619	0.88303	1.0008	1.0421	0.74811	0.81131	0.80918	0.88303	1.0008	1.0421
8.75	0.83705	0.9998	1.062	0.88206	1.0007	1.042	0.78135	0.84512	0.84393	0.88206	1.0007	1.042
10	0.83632	0.99971	1.062	0.88109	1.0005	1.0419	0.81491	0.87793	0.8776	0.88109	1.0005	1.0419
11.25	0.8369	0.99975	1.0621	0.8813	1.0004	1.0417	0.85157	0.90983	0.91013	0.8813	1.0004	1.0417
12.5	0.83557	0.99978	1.0623	0.87985	1.0003	1.0415	0.88103	0.94082	0.94183	0.87985	1.0003	1.0415
13.75	0.83665	0.99988	1.0624	0.88065	1.0001	1.0414	0.91711	0.97077	0.9726	0.88065	1.0001	1.0414
15	0.83593	1	1.0625	0.87967	1	1.0413	0.94527	1	1.0024	0.87967	1	1.0413
16.25	0.83519	1.0002	1.0626	0.8786	0.99991	1.0411	0.97315	1.0282	1.0313	0.8786	0.99991	1.0411
17.5	0.83623	1.0003	1.0627	0.87934	0.99978	1.0411	1.0049	1.0556	1.0593	0.87934	0.99978	1.0411
18.75	0.83564	1.0005	1.0627	0.87829	0.99971	1.0409	1.0302	1.082	1.0862	0.87829	0.99971	1.0409
20	0.83525	1.0005	1.0628	0.87776	0.99956	1.0408	1.0565	1.1075	1.1123	0.87776	0.99956	1.0408
21.25	0.83548	1.0006	1.0628	0.87766	0.9995	1.0407	1.0826	1.132	1.1373	0.87766	0.9995	1.0407
22.5	0.83418	1.0006	1.0629	0.87642	0.99937	1.0406	1.1056	1.1555	1.1613	0.87642	0.99937	1.0406
23.75	0.83608	1.0006	1.0629	0.87786	0.99931	1.0405	1.1325	1.1781	1.1843	0.87786	0.99931	1.0405
25	0.83561	1.0007	1.0629	0.87716	0.99918	1.0404	1.1531	1.1995	1.2063	0.87716	0.99918	1.0404
26.25	0.83268	1.0008	1.063	0.87546	0.99912	1.0404	1.1731	1.2198	1.2271	0.87546	0.99912	1.0404
27.5	0.83563	1.0008	1.063	0.87692	0.99901	1.0402	1.1952	1.2389	1.2467	0.87692	0.99901	1.0402
28.75	0.83508	1.0008	1.063	0.87634	0.99895	1.0402	1.2127	1.257	1.2651	0.87634	0.99895	1.0402
30	0.83509	1.0009	1.0631	0.87548	0.99884	1.0401	1.2285	1.2736	1.2821	0.87548	0.99884	1.0401
32	0.83197	1.0009	1.0631	0.87311	0.99877	1.04	1.2517	1.2975	1.3064	0.87311	0.99877	1.04
33	0.83607	1.001	1.0631	0.87782	0.99874	1.04	1.2741	1.3081	1.3173	0.87782	0.99874	1.04
34	0.83332	1.001	1.0631	0.8724	0.99871	1.0399	1.2711	1.3176	1.327	0.8724	0.99871	1.0399
35	0.8366	1.0009	1.0631	0.87816	0.99863	1.0399	1.294	1.3261	1.3357	0.87816	0.99863	1.0399
36	0.82669	1.001	1.0632	0.86695	0.99861	1.0398	1.2829	1.3333	1.3431	0.86695	0.99861	1.0398
37	0.84971	1.001	1.0632	0.88567	0.99859	1.0398	1.3101	1.3394	1.3493	0.88567	0.99859	1.0398
38	0.73357	1.001	1.0632	0.82624	0.99858	1.0398	1.2548	1.3444	1.3544	0.82624	0.99858	1.0398
39	0.78815	1.001	1.0632	0.84161	0.99856	1.0398	1.2454	1.3479	1.358	0.84161	0.99856	1.0398
40	0.71604	1.001	1.0632	0.82091	0.99857	1.0398	0.83365	1.35	1.3601	0.82091	0.99857	1.0398

Table A4. Performance as a function of HTF temperature (MOD1) table for Partial Cooling cycle. "Low", "On" and "High" respectively refer to the three normalised mass flows employed: 0.2, 1 and 1.05.

T_{HTF}	$\dot{W}_{cycle,low}$	$\dot{W}_{cycle,on}$	$\dot{W}_{cycle,high}$	$Q_{in,low}$	$Q_{in,on}$	$Q_{in,high}$	$\dot{W}_{cooling,low}$	$\dot{W}_{cooling,on}$	$\dot{W}_{cooling,high}$	$\dot{m}_{w,cooling,low}$	$\dot{m}_{w,cooling on}$	$\dot{m}_{w,cooling,high}$
700	0.13893	0.85703	0.912	0.16588	0.91658	0.97042	0.088139	1	0.90031	0.16588	0.91658	0.97042
705	0.14061	0.86785	0.92333	0.1666	0.92329	0.97745	0.091209	1	0.90031	0.1666	0.92329	0.97745
710	0.14236	0.87873	0.9345	0.16742	0.9299	0.98448	0.097024	1	0.90031	0.16742	0.9299	0.98448
715	0.14399	0.8888	0.94589	0.16815	0.93606	0.9916	0.10127	1	0.90031	0.16815	0.93606	0.9916
720	0.14566	0.89978	0.9572	0.16895	0.9429	0.9986	0.10724	1	0.90031	0.16895	0.9429	0.9986
725	0.1474	0.91036	0.96783	0.16975	0.94992	1.0052	0.1124	1	0.90031	0.16975	0.94992	1.0052
730	0.14859	0.9238	0.97924	0.17056	0.95517	1.0121	0.11938	1	0.90031	0.17056	0.95517	1.0121
735	0.14928	0.93018	0.98974	0.1709	0.96188	1.0186	0.11938	1	0.90031	0.1709	0.96188	1.0186
740	0.15066	0.94373	1.0005	0.17201	0.96718	1.0258	0.11938	1	0.90031	0.17201	0.96718	1.0258
745	0.15375	0.95011	1.0143	0.17308	0.97367	1.0314	0.14248	1	0.90031	0.17308	0.97367	1.0314
750	0.15475	0.96253	1.0212	0.17384	0.97817	1.0383	0.14248	1	0.90031	0.17384	0.97817	1.0383
755	0.15637	0.96885	1.0346	0.17451	0.98478	1.0437	0.14463	1	0.90031	0.17451	0.98478	1.0437
760	0.15821	0.9818	1.0413	0.17511	0.98953	1.0505	0.15914	1	0.90031	0.17511	0.98953	1.0505
765	0.1597	0.98829	1.0541	0.17586	0.99594	1.0552	0.16548	1	0.90031	0.17586	0.99594	1.0552
770	0.16117	1	1.0608	0.17659	1	1.062	0.1725	1	0.90031	0.17659	1	1.062
775	0.16246	1.007	1.0731	0.17729	1.0065	1.0666	0.17775	1	0.90031	0.17729	1.0065	1.0666
780	0.16361	1.0178	1.08	0.1778	1.01	1.0736	0.17775	1	0.90031	0.1778	1.01	1.0736
785	0.16461	1.025	1.0929	0.17828	1.0167	1.0783	0.17775	1	0.90031	0.17828	1.0167	1.0783
790	0.16558	1.0364	1.1003	0.17876	1.0209	1.085	0.17775	0.99727	0.90031	0.17876	1.0209	1.085
795	0.16736	1.0429	1.1111	0.17948	1.0241	1.0886	0.17775	0.98552	0.88922	0.17948	1.0241	1.0886
800	0.16773	1.0491	1.1172	0.18055	1.0308	1.0956	0.17992	0.9859	0.88824	0.18055	1.0308	1.0956

Table A5. Performance as a function of HTF mass flow rate (MOD2) table for Partial Cooling cycle. "Low", "On" and "High" respectively refer to the three values of ambient temperature employed: 5, 15 and 40 °C.

T_{HTF}	$\dot{W}_{cycle,low}$	$\dot{W}_{cycle,on}$	$\dot{W}_{cycle,high}$	$Q_{in,low}$	$Q_{in,on}$	$Q_{in,high}$	$\dot{W}_{cooling,low}$	$\dot{W}_{cooling,on}$	$\dot{W}_{cooling,high}$	$\dot{m}_{w,cooling,low}$	$\dot{m}_{w,cooling,on}$	$\dot{m}_{w,cooling,high}$
0.2	0.097547	0.079874	0.092687	0.1863	0.16089	0.16269	0.051009	0.25729	0.7088	0.1863	0.16089	0.16269
0.21	0.10537	0.079874	0.092687	0.19425	0.16986	0.17154	0.051009	0.25729	0.7088	0.19425	0.16986	0.17154
0.22	0.11327	0.079874	0.092687	0.20226	0.17888	0.18042	0.051009	0.25729	0.7088	0.20226	0.17888	0.18042
0.23	0.12126	0.079874	0.092687	0.21032	0.18793	0.18934	0.051009	0.25729	0.7088	0.21032	0.18793	0.18934
0.24	0.12934	0.079874	0.092687	0.21843	0.19702	0.1983	0.051009	0.25729	0.7088	0.21843	0.19702	0.1983
0.25	0.1375	0.079874	0.092687	0.2266	0.20614	0.2073	0.051009	0.25729	0.7088	0.2266	0.20614	0.2073
0.26	0.14574	0.079874	0.092687	0.23483	0.21531	0.21633	0.051009	0.25729	0.7088	0.23483	0.21531	0.21633
0.27	0.15407	0.079874	0.092687	0.24311	0.22451	0.22541	0.051009	0.25729	0.7088	0.24311	0.22451	0.22541
0.28	0.16249	0.079874	0.092687	0.25145	0.23375	0.23453	0.051009	0.25729	0.7088	0.25145	0.23375	0.23453
0.29	0.17099	0.079874	0.092687	0.25984	0.24303	0.24368	0.051009	0.25729	0.7088	0.25984	0.24303	0.24368
0.3	0.17958	0.079874	0.092687	0.26829	0.25088	0.25149	0.051009	0.25729	0.7088	0.26829	0.25088	0.25149
0.31	0.18825	0.11541	0.12263	0.27679	0.26124	0.26155	0.051009	0.25729	0.7088	0.27679	0.26124	0.26155
0.32	0.197	0.13891	0.14326	0.28535	0.27111	0.27116	0.051009	0.25798	0.71059	0.28535	0.27111	0.27116
0.33	0.20585	0.15708	0.15987	0.29397	0.2807	0.28055	0.051009	0.26077	0.71059	0.29397	0.2807	0.28055
0.34	0.21477	0.1728	0.17468	0.30264	0.29023	0.28989	0.051009	0.26662	0.71428	0.30264	0.29023	0.28989
0.35	0.22378	0.18693	0.18799	0.31136	0.29969	0.29927	0.051009	0.27197	0.71428	0.31136	0.29969	0.29927
0.36	0.23288	0.2004	0.20133	0.32014	0.30923	0.30882	0.051009	0.27833	0.72156	0.32014	0.30923	0.30882
0.37	0.24206	0.21266	0.21348	0.32898	0.31852	0.31811	0.051009	0.27992	0.72344	0.32898	0.31852	0.31811
0.38	0.25133	0.22631	0.2258	0.33787	0.32851	0.32758	0.051009	0.29775	0.72985	0.33787	0.32851	0.32758
0.39	0.26068	0.2382	0.23745	0.34681	0.3379	0.33692	0.051009	0.30287	0.73172	0.34681	0.3379	0.33692
0.4	0.27012	0.25072	0.25003	0.35581	0.34768	0.34675	0.051009	0.3137	0.74474	0.35581	0.34768	0.34675
0.41	0.27964	0.2623	0.26172	0.36487	0.35713	0.35635	0.051009	0.31729	0.74664	0.36487	0.35713	0.35635
0.42	0.28925	0.275	0.27366	0.37398	0.36717	0.3659	0.051009	0.33027	0.75616	0.37398	0.36717	0.3659
0.43	0.29895	0.28647	0.28524	0.38315	0.37674	0.37556	0.051009	0.33306	0.75807	0.38315	0.37674	0.37556
0.44	0.30872	0.29914	0.29778	0.39237	0.38676	0.38551	0.051009	0.34792	0.77274	0.39237	0.38676	0.38551
0.45	0.31859	0.31073	0.30921	0.40165	0.3964	0.3952	0.051009	0.35083	0.77274	0.40165	0.3964	0.3952
0.46	0.32854	0.32318	0.32152	0.41098	0.40637	0.40498	0.051009	0.36317	0.78465	0.41098	0.40637	0.40498
0.47	0.33857	0.33381	0.3316	0.42037	0.41603	0.41475	0.051009	0.36317	0.78465	0.42037	0.41603	0.41475
0.48	0.34869	0.34628	0.34378	0.42982	0.42637	0.42502	0.051009	0.38896	0.81314	0.42982	0.42637	0.42502
0.49	0.35889	0.35578	0.35356	0.43931	0.43621	0.43486	0.051009	0.38999	0.8151	0.43931	0.43621	0.43486
0.5	0.36918	0.3679	0.36557	0.44887	0.44626	0.44526	0.051009	0.40888	0.83707	0.44887	0.44626	0.44526
0.51	0.37956	0.37751	0.3762	0.45848	0.45608	0.45517	0.051009	0.40888	0.84612	0.45848	0.45608	0.45517
0.52	0.38452	0.39074	0.38727	0.46536	0.46691	0.46514	0.051009	0.43679	0.85852	0.46536	0.46691	0.46514
0.53	0.39569	0.40075	0.39791	0.47555	0.47682	0.47486	0.065572	0.43788	0.85852	0.47555	0.47682	0.47486
0.54	0.40661	0.41315	0.40804	0.48514	0.48712	0.48481	0.073752	0.45509	0.85852	0.48514	0.48712	0.48481
0.55	0.41664	0.42342	0.42237	0.49509	0.49719	0.49599	0.073752	0.45509	0.91197	0.49509	0.49719	0.49599
0.56	0.43122	0.43747	0.43273	0.5064	0.50838	0.50609	0.11457	0.49046	0.91197	0.5064	0.50838	0.50609
0.57	0.44139	0.44855	0.44463	0.51635	0.51874	0.51617	0.11762	0.49497	0.92518	0.51635	0.51874	0.51617
0.58	0.45373	0.46076	0.45538	0.52652	0.52923	0.52642	0.13143	0.51121	0.92518	0.52652	0.52923	0.52642
0.59	0.46424	0.4724	0.4698	0.53682	0.53938	0.53785	0.13143	0.51666	0.97638	0.53682	0.53938	0.53785
0.6	0.47912	0.48309	0.48047	0.54827	0.5495	0.54796	0.16843	0.51666	0.97638	0.54827	0.5495	0.54796
0.61	0.48967	0.4979	0.49395	0.55845	0.56114	0.55925	0.16984	0.55624	1.0128	0.55845	0.56114	0.55925
0.62	0.50246	0.50912	0.50517	0.56886	0.5716	0.56897	0.18584	0.55905	1.0175	0.56886	0.5716	0.56897
0.63	0.51321	0.52125	0.5164	0.5791	0.58177	0.57948	0.18584	0.56851	1.0175	0.5791	0.58177	0.57948
0.64	0.52738	0.53263	0.53045	0.59021	0.59251	0.5906	0.21446	0.56851	1.0757	0.59021	0.59251	0.5906
0.65	0.53779	0.54647	0.54087	0.60054	0.60341	0.6011	0.21446	0.60628	1.0757	0.60054	0.60341	0.6011
0.66	0.55101	0.55689	0.55434	0.61144	0.61395	0.61238	0.23966	0.60628	1.1447	0.61144	0.61395	0.61238
0.67	0.56133	0.56959	0.56503	0.62203	0.62464	0.62242	0.23966	0.62862	1.1447	0.62203	0.62464	0.62242
0.68	0.57414	0.58025	0.57589	0.63251	0.63541	0.63275	0.26304	0.62862	1.1447	0.63251	0.63541	0.63275
0.69	0.58469	0.59335	0.58706	0.6432	0.64602	0.64312	0.26304	0.65898	1.1447	0.6432	0.64602	0.64312
0.7	0.59777	0.60434	0.59782	0.6539	0.6571	0.6534	0.28504	0.65898	1.1447	0.6539	0.6571	0.6534
0.71	0.60848	0.61753	0.60912	0.6648	0.66792	0.66365	0.28504	0.68777	1.1447	0.6648	0.66792	0.66365
0.72	0.6209	0.62832	0.62004	0.6754	0.67874	0.67421	0.29829	0.68777	1.1447	0.6754	0.67874	0.67421
0.73	0.63212	0.64132	0.63158	0.68611	0.68948	0.68446	0.30119	0.70993	1.1447	0.68611	0.68948	0.68446
0.74	0.64477	0.65246	0.64264	0.69668	0.70052	0.6954	0.31102	0.71109	1.1447	0.69668	0.70052	0.6954
0.75	0.65606	0.66551	0.6546	0.70749	0.71109	0.70596	0.31334	0.72916	1.1779	0.70749	0.71109	0.70596
0.76	0.66882	0.67686	0.66588	0.7181	0.72222	0.71697	0.32135	0.73036	1.1799	0.7181	0.72222	0.71697
0.77	0.68077	0.69035	0.67906	0.72901	0.73287	0.72784	0.32546	0.74783	1.2274	0.72901	0.73287	0.72784
0.78	0.69306	0.70177	0.69058	0.7397	0.74433	0.73902	0.33052	0.74783	1.2295	0.7397	0.74433	0.73902
0.79	0.70498	0.71596	0.70396	0.75064	0.75529	0.75014	0.33482	0.76442	1.2731	0.75064	0.75529	0.75014
0.8	0.71693	0.72749	0.71539	0.76142	0.76645	0.76121	0.33921	0.76573	1.2753	0.76142	0.76645	0.76121
0.81	0.72927	0.7413	0.72921	0.77263	0.77768	0.77235	0.34278	0.77747	1.3189	0.77263	0.77768	0.77235
0.82	0.74105	0.75345	0.74116	0.78326	0.78912	0.78368	0.34642	0.78006	1.3236	0.78326	0.78912	0.78368
0.83	0.75368	0.76673	0.75457	0.79461	0.7999	0.79457	0.34918	0.78933	1.367	0.79461	0.7999	0.79457
0.84	0.76585	0.77944	0.76673	0.80533	0.81141	0.80601	0.34918	0.7933	1.3719	0.80533	0.81141	0.80601
0.85	0.77814	0.7925	0.78031	0.81644	0.8225	0.81696	0.34918	0.80001	1.4124	0.81644	0.8225	0.81696
0.86	0.79074	0.80531	0.79286	0.82734	0.83383	0.82853	0.34918	0.8054	1.4227	0.82734	0.83383	0.82853
0.87	0.80373	0.81824	0.80642	0.83847	0.84521	0.83954	0.34918	0.80945	1.4571	0.83847	0.84521	0.83954
0.88	0.81686	0.83106	0.81937	0.84983	0.85631	0.8511	0.34918	0.81489	1.4705	0.84983	0.85631	0.8511
0.89	0.83055	0.84409	0.8329	0.86115	0.8679	0.86247	0.34918	0.81762	1.4952	0.86115	0.8679	0.86247
0.9	0.84324	0.85747	0.84604	0.87261	0.87912	0.87381	0.34918	0.82036	1.4952	0.87261	0.87912	0.87381
0.91	0.85559	0.87128	0.85956	0.88266	0.89101	0.88568	0.34918	0.82036	1.4952	0.88266	0.89101	0.88568
0.92	0.87609	0.88457	0.87257	0.89805	0.90216	0.8967	0.26777	0.82036	1.4952	0.89805	0.90216	0.8967
0.93	0.88932	0.89804	0.88655	0.90989	0.91397	0.90864	0.27114	0.82036	1.4952	0.90989	0.91397	0.90864
0.94	0.90266	0.91102	0.90006	0.92115	0.92537	0.92023	0.27529	0.82036	1.4952	0.92115	0.92537	0.92023
0.95	0.91623	0.92419	0.91379	0.93305	0.9371	0.9315	0.27836	0.82036	1.4952	0.93305	0.9371	0.9315
0.96	0.92999	0.93772	0.92776	0.94467	0.94829	0.94347	0.28245	0.82036	1.4952	0.94467	0.94829	0.94347
0.97	0.9449	0.95197	0.95198	0.95674	0.96029	0.95907	0.28545	0.82036	1.8993	0.95674	0.96029	0.95907
0.98	0.97427	0.97445	0.96484	0.97576	0.97593	0.97116	0.4914	1.0181	1.8748	0.97576	0.97593	0.97116
0.99	0.98715	0.98701	0.97739	0.98781	0.98788	0.98304	0.49489	1.0097	1.8453	0.98781	0.98788	0.98304
1	1.0001	1	0.98994	1.0002	1	0.99505	0.49849	1	1.8102	1.0002	1	0.99505
1.01	1.0129	1.0132	1.0025	1.0121	1.0126	1.0074	0.50288	0.9883	1.7701	1.0121	1.0126	1.0074
1.02	1.026	1.0258	1.0145	1.0248	1.0245	1.0189	0.50676	0.97835	1.729	1.0248	1.0245	1.0189
1.03	1.0387	1.0386	1.0272	1.0369	1.0371	1.0312	0.51008	0.9647	1.6814	1.0369	1.0371	1.0312
1.04	1.0516	1.0518	1.0392	1.0491	1.0493	1.0431	0.51008	0.96298	1.6335	1.0491	1.0493	1.0431
1.05	1.065	1.0654	1.0514	1.062	1.0623	1.055	0.51124	0.96298	1.5862	1.062	1.0623	1.055

Table A6. Performance as a function of ambient temperature (MOD3) table for Partial Cooling cycle. "Low", "On" and "High" respectively refer to the three values of molten salt (hot) temperature levels employed: 700, 770 and 800 °C.

T_{HTF}	$\dot{W}_{cycle,low}$	$\dot{W}_{cycle,on}$	$\dot{W}_{cycle,high}$	$Q_{in,low}$	$Q_{in,on}$	$Q_{in,high}$	$\dot{W}_{cooling,low}$	$\dot{W}_{cooling,on}$	$\dot{W}_{cooling,high}$	$\dot{m}_{w,cooling,low}$	$\dot{m}_{w,cooling on}$	$\dot{m}_{w,cooling,high}$
5	0.87197	1.0106	1.051	0.92305	1.0058	1.0289	0.52313	0.54897	0.5307	0.92305	1.0058	1.0289
6.25	0.86862	1.0157	1.056	0.92122	1.0057	1.0287	0.55291	0.58798	0.56835	0.92122	1.0057	1.0287
7.5	0.86503	1.0139	1.0541	0.91958	1.0046	1.0277	0.59877	0.6321	0.6104	0.91958	1.0046	1.0277
8.75	0.86266	1.0107	1.0509	0.91818	1.0031	1.0261	0.64616	0.67912	0.65462	0.91818	1.0031	1.0261
10	0.86076	1.0078	1.0479	0.9173	1.0016	1.0247	0.70367	0.72727	0.70019	0.9173	1.0016	1.0247
11.25	0.86147	1.003	1.0424	0.91735	1.0006	1.0236	0.77223	0.78855	0.75934	0.91735	1.0006	1.0236
12.5	0.86147	0.99992	1.0401	0.91736	0.99998	1.0231	0.8467	0.85827	0.82408	0.91736	0.99998	1.0231
13.75	0.8613	1.0002	1.0401	0.91736	0.99999	1.0231	0.92843	0.92682	0.88976	0.91736	0.99999	1.0231
15	0.86194	1	1.0404	0.9174	1	1.0231	1.0162	1	0.95711	0.9174	1	1.0231
16.25	0.86138	1.0003	1.0402	0.91738	1	1.0231	1.1199	1.075	1.029	0.91738	1	1.0231
17.5	0.86192	1.0005	1.0406	0.91741	1	1.0231	1.2319	1.1526	1.1025	0.91741	1	1.0231
18.75	0.86189	1.0005	1.0402	0.9174	1	1.0231	1.3482	1.2329	1.1821	0.9174	1	1.0231
20	0.86152	1.0001	1.0403	0.9174	1	1.0231	1.454	1.3145	1.2621	0.9174	1	1.0231
21.25	0.86232	1.0007	1.0405	0.91743	1.0001	1.0231	1.5404	1.3904	1.3425	0.91743	1.0001	1.0231
22.5	0.86182	1.0003	1.0407	0.91741	1	1.0232	1.6166	1.4649	1.4205	0.91741	1	1.0232
23.75	0.86155	1.0006	1.0409	0.91741	1.0001	1.0231	1.6825	1.5306	1.4934	0.91741	1.0001	1.0231
25	0.86202	1.0008	1.0405	0.91745	1.0001	1.0231	1.7397	1.5917	1.5614	0.91745	1.0001	1.0231
26.25	0.86236	1.0005	1.0402	0.91746	1.0001	1.0231	1.7907	1.6479	1.6216	0.91746	1.0001	1.0231
27.5	0.86204	1.0008	1.0407	0.91744	1.0001	1.0232	1.837	1.6969	1.6745	0.91744	1.0001	1.0232
28.75	0.86178	1.0005	1.041	0.91745	1.0001	1.0232	1.8788	1.743	1.7216	0.91745	1.0001	1.0232
30	0.86238	1.0009	1.04	0.91749	1.0001	1.0231	1.9153	1.7819	1.7671	0.91749	1.0001	1.0231
32	0.86256	1.0004	1.0413	0.91749	1.0001	1.0232	1.9657	1.8383	1.8232	0.91749	1.0001	1.0232
33	0.86186	1.0005	1.0404	0.91747	1.0001	1.0232	1.9889	1.8623	1.8496	0.91747	1.0001	1.0232
34	0.86263	1.0008	1.0412	0.9175	1.0001	1.0232	2.0074	1.8832	1.8705	0.9175	1.0001	1.0232
35	0.86232	1.0009	1.0408	0.91749	1.0001	1.0232	2.0248	1.901	1.8904	0.91749	1.0001	1.0232
36	0.86216	1.0007	1.0407	0.91749	1.0001	1.0232	2.0396	1.9177	1.907	0.91749	1.0001	1.0232
37	0.8628	1.0013	1.0413	0.91751	1.0001	1.0232	2.051	1.929	1.9193	0.91751	1.0001	1.0232
38	0.86247	1.0002	1.041	0.91748	1.0001	1.0232	2.0601	1.9414	1.9297	0.91748	1.0001	1.0232
39	0.8622	1.0017	1.0407	0.91747	1.0003	1.0232	2.0671	1.9485	1.938	0.91747	1.0003	1.0232
40	0.86193	1.0016	1.0404	0.91748	1.0003	1.0232	2.0719	1.951	1.9433	0.91748	1.0003	1.0232

References

1. Office of Energy Efficiency & Renewable Energy—The Sunshot Iniciative. Available online: https://www.energy.gov/eere/solar/sunshot-initiative (accessed on 15 March 2019).

2. Manzolini, G.; Binotti, M.; Bonalumi, D.; Invernizzi, C.; Iora, P. CO2 mixtures as innovative working fluid in power cycles applied to solar plants. Techno-economic assessment. *Solar Energy* **2019**, *181*, 530–544. [CrossRef]

3. Binotti, M.; Di Marcoberardino, G.; Iora, P.; Invernizzi, C.M.; Manzolini, G. Supercritical carbon dioxide/alternative fluids blends for efficiency upgrade of solar power plant. In Proceedings of the 3rd European Conference on Supercritical CO$_2$ (sCO$_2$) Power Systems 2019, Paris, France, 19–20 September 2019; pp. 222–229.

4. Crespi, F.; Gavagnin, G.; Sánchez, D.; Martínez, G. Supercritical Carbon Dioxide Cycles for Power Generation: A Review. *Appl. Energy* **2017**, *195*, 152–183. [CrossRef]

5. Sulzer, G. Verfahren zur Erzeugung von Arbeit aus Warme. *Swiss Patent* **1950**, 269599.

6. Angelino, G. Carbon Dioxide Condensation Cycles for Power Production. *J. Eng. Power* **1968**, *90*, 287–295. [CrossRef]

7. Feher, E. The supercritical thermodynamic power cycle. *Energy Conv. Manag.* **1968**, *8*, 85–90. [CrossRef]

8. Crespi, F.; Gavagnin, G.; Sánchez, D.; Martínez, G. Analysis of the Thermodynamic Potential of Supercritical Carbon Dioxide Cycles: A Systematic Approach. *J. Eng. Gas Turb. Power* **2017**, *140*, 051701. [CrossRef]

9. Crespi, F.; Sánchez, D.; Rodríguez, J.; Gavagnin, G. Fundamental Thermo-Economic Approach to Selecting sCO$_2$ Power Cycles for CSP Applications. *Energy Procedia* **2017**, *129*, 963–970. [CrossRef]

10. Crespi, F.; Sánchez, D.; Sánchez, T.; Martínez, G.S. Capital Cost Assessment of Concentrated Solar Power Plants Based on Supercritical Carbon Dioxide Power Cycles. *J. Eng. Gas Turb. Power* **2019**, *141*. [CrossRef]

11. Crespi, F.; Sánchez, D.; Sánchez-Lencero, T.; Martínez, G.; Muñoz, A. Off-design operation of Supercritical Carbon Dioxide Power Cycles in Concentrated Solar Power plants. *Appl. Thermal Eng.* **2020**, in press.

12. Various. *Report on Best Available Technologies (BAT) for Central Receiver Systems*; Technical Report; Abengoa Energia, University of Seville: Seville, Spain, 2019.

13. Martin, M. Techno-Economic Assessment of Concentrated Solar Power Tower Plants Integrating Pressurised Air Rceivers and Gas Turbines. Ph.D. Thesis, University of Seville, Seville, Spain, 2018. (In Spanish)

14. Williams, D. *Assessment of Candidate Molten Salt Coolants for the NGNP/NHI Heat-Transfer Loop*; Technical Report; Oak Ridge National Lab (ORNL): Oak Ridge, TN, USA, 2006.

15. Mehos, M.; Turchi, C.; Vidal, J.; Wagner, M.; Ma, Z.; Ho, C.; Kolb, W.; Andraka, C.; Kruizenga, A. *Concentrating Solar Power Gen3 Demonstration Roadmap*; Technical Report; National Renewable Energy Lab (NREL): Golden, CO, USA, 2017.

16. Yoon, S.; Sabharwall, P.; Kim, E. *Analytical Study on Thermal and Mechanical Design of Printed Circuit Heat Exchanger*; Technical Report; Idaho Nation Laboratory: Idaho Falls, ID, USA, 2013.

17. Prieto, C.; Fereres, S.; Ruiz-Cabañas, F.J.; Rodriguez-Sanchez, A.; Montero, C. Carbonate molten salt solar thermal pilot facility: Plant design, commissioning and operation up to 700 °C. *Renew. Energy* **2020**, *151*, 528–541. [CrossRef]

18. Turchi, C.S.; Ma, Z.; Neises, T.W.; Wagner, M.J. Thermodynamic study of advanced supercritical carbon dioxide power cycles for concentrating solar power systems. *J. Solar Energy Eng.* **2013**, *135*, 041007.

19. Dyreby, J.; Klein, S.; Nellis, G.; Reindl, D. Design considerations for supercritical carbon dioxide Brayton cycles with recompression. *J. Eng. Gas Turb. Power* **2014**, *136*, 101701. [CrossRef]

20. Silvestri, G. *Eddystone Station, 325 MW Generating Unit, A Brief History*; ASME: Philadelphia, PA, USA, 2003.

21. Weiland, N.T.; Lance, B.W.; Pidaparti, S.R. sCO$_2$ Power Cycle Component Cost Correlations From DOE Data Spanning Multiple Scales and Applications. In Proceedings of the ASME Turbo Expo 2019: Turbomachinery Technical Conference and Exposition, Phoenix, AZ, USA, 17–21 June 2019.

22. Carlson, M.D.; Middleton, B.M.; Ho, C.K. Techno-economic comparison of solar-driven SCO$_2$ Brayton cycles using component cost models baselined with vendor data and estimates. In *Proceedings of the ASME 2017—11th International Conference on Energy Sustainability Collocated with the ASME 2017 Power Conference Joint With ICOPE-17, the ASME 2017 15th International Conference on Fuel Cell Science, Engineering and Technology, and the ASME 2017 Nuclear Forum (Digital Collection)*; American Society of Mechanical Engineers (ASME): New York, NY, USA, 2017.

23. IRENA. *Renewable Power Generation Costs in 2018*; International Renewable Energy Agency: Abu Dhabi, UAE, 2019.

24. Allam, R.; Fetvedt, J.; Forrest, B.; Freed, D. The Oxy-Fuel, Supercritical CO$_2$ Allam Cycle: New Cycle Developments to Produce Even Lower-Cost Electricity From Fossil Fuels Without Atmospheric Emissions. In *ASME Turbo Expo 2014: Turbine Technical Conference and Exposition*; ASME: New York, NY, USA, 2014.

25. Kulhánek, M.; Dostal, V. Thermodynamic analysis and comparison of supercritical carbon dioxide cycles. In Proceedings of the Supercritical CO$_2$ Power Cycle Symposium, Boulder, CO, USA, 24–25 May 2011; pp. 1–7.

26. Neises, T.; Turchi, C. Supercritical CO$_2$ Power Cycles: Design Considerations for Concentrating Solar Power. In Proceedings of the 4th Supercritical CO$_2$ Power Cycles Symposium, Pittsburgh, PA, USA, 9–10 September 2014; Volume 2, pp. 9–10.

27. Crespi, F. Thermo-Economic Assessment of Supercritical CO$_2$ Power Cycles for Concentrated Solar Power Plants. Ph.D. Thesis, Unviersity of Seville, Seville, Spain, 2019.

28. Schmitt, J.; Wilkes, J.; Allison, T.; Bennett, J.; Wygant, K.; Pelton, R. Lowering the Levelized Cost of Electricity of a Concentrating Solar Power Tower With a Supercritical Carbon Dioxide Power Cycle. In *ASME Turbo Expo 2017: Turbomachinery Technical Conference and Exposition (Digital Collection)*; American Society of Mechanical Engineers (ASME): New York, NY, USA, 2017.

29. Ho, C.; Mehos, M.; Turchi, C.; Wagner, M. Probabilistic Analysis of Power Tower Systems to Achieve Sunshot Goals. *Energy Procedia* **2014**, *49*, 1410–1419. [CrossRef]

30. Crespi, F.; Sánchez, D.; Hoopes, K.; Choi, B.; Kuek, N. The Conductance Ratio Method for Off-Design Heat Exchanger Modeling and its Impact on an sCO$_2$ Recompression Cycle. In *ASME Turbo Expo 2019: Turbomachinery Technical Conference and Exposition (Digital Collection)*; American Society of Mechanical Engineers (ASME): New York, NY, USA, 2019.

31. Sánchez, D.; Bortkiewicz, A.; Rodr'íguez, J.; Martínez, G.; Gavagnin, G.; Sánchez, T. A methodology to identify potential markets for small-scale solar thermal power generators. *Appl. Energy* **2016**, *169*, 287–300. [CrossRef]

MDPI
St. Alban-Anlage 66
4052 Basel
Switzerland
Tel. +41 61 683 77 34
Fax +41 61 302 89 18
www.mdpi.com

Applied Sciences Editorial Office
E-mail: applsci@mdpi.com
www.mdpi.com/journal/applsci